These safety symbols are used in laboratory and field investigations in this book to indicate possible hazards. Learn the meaning of each symbol and refer to this page often. *Remember to wash your hands thoroughly after completing lab procedures.*

PROTECTIVE EQUIPMENT Do not begin any lab without the proper protection equipment.

GOGGLES Proper eye protection must be worn when performing or observing science activities that involve items or conditions as listed below.	**APRON** Wear an approved apron when using substances that could stain, wet, or destroy cloth.	**SOAP** Wash hands with soap and water before removing goggles and after all lab activities.	**GLOVES** Wear gloves when working with biological materials, chemicals, animals, or materials that can stain or irritate hands.

LABORATORY HAZARDS

Symbols	Potential Hazards	Precaution	Response
DISPOSAL	contamination of classroom or environment due to improper disposal of materials such as chemicals and live specimens	• DO NOT dispose of hazardous materials in the sink or trash can. • Dispose of wastes as directed by your teacher.	• If hazardous materials are disposed of improperly, notify your teacher immediately.
EXTREME TEMPERATURE	skin burns due to extremely hot or cold materials such as hot glass, liquids, or metals; liquid nitrogen; dry ice	• Use proper protective equipment, such as hot mitts and/or tongs, when handling objects with extreme temperatures.	• If injury occurs, notify your teacher immediately.
SHARP OBJECTS	punctures or cuts from sharp objects such as razor blades, pins, scalpels, and broken glass	• Handle glassware carefully to avoid breakage. • Walk with sharp objects pointed downward, away from you and others.	• If broken glass or injury occurs, notify your teacher immediately.
ELECTRICAL	electric shock or skin burn due to improper grounding, short circuits, liquid spills, or exposed wires	• Check condition of wires and apparatus for fraying or uninsulated wires, and broken or cracked equipment. • Use only GFCI-protected outlets	• DO NOT attempt to fix electrical problems. Notify your teacher immediately.
CHEMICAL	skin irritation or burns, breathing difficulty, and/or poisoning due to touching, swallowing, or inhalation of chemicals such as acids, bases, bleach, metal compounds, iodine, poinsettias, pollen, ammonia, acetone, nail polish remover, heated chemicals, mothballs, and any other chemicals labeled or known to be dangerous	• Wear proper protective equipment such as goggles, apron, and gloves when using chemicals. • Ensure proper room ventilation or use a fume hood when using materials that produce fumes. • NEVER smell fumes directly. • NEVER taste or eat any material in the laboratory.	• If contact occurs, immediately flush affected area with water and notify your teacher. • If a spill occurs, leave the area immediately and notify your teacher.
FLAMMABLE	unexpected fire due to liquids or gases that ignite easily such as rubbing alcohol	• Avoid open flames, sparks, or heat when flammable liquids are present.	• If a fire occurs, leave the area immediately and notify your teacher.
OPEN FLAME	burns or fire due to open flame from matches, Bunsen burners, or burning materials	• Tie back loose hair and clothing. • Keep flame away from all materials. • Follow teacher instructions when lighting and extinguishing flames. • Use proper protection, such as hot mitts or when handling hot objects.	
ANIMAL SAFETY	injury to or from laboratory animals	• Wear proper protective equipment such as gloves, apron, and goggles when working with animals. • Wash hands after handling animals.	teacher immediately.
BIOLOGICAL	infection or adverse reaction due to contact with organisms such as bacteria, fungi, and biological materials such as blood, animal or plant materials	• Wear proper protective equipment such as gloves, goggles, and apron when working with biological materials. • Avoid skin contact with an organism or any part of the organism. • Wash hands after handling organisms.	• If contact occurs, wash the affected area and notify your teacher immediately.
FUME	breathing difficulties from inhalation of fumes from substances such as ammonia, acetone, nail polish remover, heated chemicals, and mothballs	• Wear goggles, apron, and gloves. • Ensure proper room ventilation or use a fume hood when using substances that produce fumes. • NEVER smell fumes directly.	• If a spill occurs, leave area and notify your teacher immediately.
IRRITANT	irritation of skin, mucous membranes, or respiratory tract due to materials such as acids, bases, bleach, pollen, mothballs, steel wool, and potassium permanganate	• Wear goggles, apron, and gloves. • Wear a dust mask to protect against fine particles.	• If skin contact occurs, immediately flush the affected area with water and notify your teacher.
RADIOACTIVE	excessive exposure from alpha, beta, and gamma particles	• Remove gloves and wash hands with soap and water before removing remainder of protective equipment.	• If cracks or holes are found in the container, notify your teacher immediately.

Your online portal to everything you need

connectED.mcgraw-hill.com

Look for these icons to access exciting digital resources

 Video

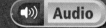 Audio

 Review

 Inquiry

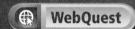 WebQuest

 Assessment

Concepts in Motion

McGraw Hill Education

ATOMS AND ELEMENTS

iSCIENCE

Glencoe

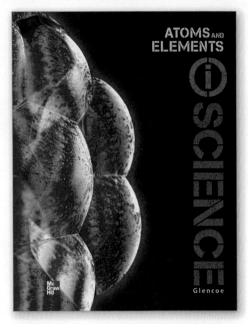

Bubbles

The iridescent colors of these soap bubbles result from a property called interference. Light waves reflect off both outside and inside surfaces of bubbles. When this happens, the waves interfere with each other and you see different colors. The thickness of the soap film that forms a bubble also affects interference.

The McGraw·Hill Companies

 Education

Copyright © 2012 The McGraw-Hill Companies, Inc. All rights reserved. No part of this publication may be reproduced or distributed in any form or by any means, or stored in a database or retrieval system, without the prior written consent of The McGraw-Hill Companies, Inc., including, but not limited to, network storage or transmission, or broadcast for distance learning.

Send all inquiries to:
McGraw-Hill Education
8787 Orion Place
Columbus, OH 43240-4027

ISBN: 978-0-07-888021-6
MHID: 0-07-888021-1

Printed in the United States of America.

2 3 4 5 6 7 8 9 10 11 DOW 15 14 13 12 11

Authors and Contributors

Authors

American Museum of Natural History
New York, NY

Michelle Anderson, MS
Lecturer
The Ohio State University
Columbus, OH

Juli Berwald, PhD
Science Writer
Austin, TX

John F. Bolzan, PhD
Science Writer
Columbus, OH

Rachel Clark, MS
Science Writer
Moscow, ID

Patricia Craig, MS
Science Writer
Bozeman, MT

Randall Frost, PhD
Science Writer
Pleasanton, CA

Lisa S. Gardiner, PhD
Science Writer
Denver, CO

Jennifer Gonya, PhD
The Ohio State University
Columbus, OH

Mary Ann Grobbel, MD
Science Writer
Grand Rapids, MI

Whitney Crispen Hagins, MA, MAT
Biology Teacher
Lexington High School
Lexington, MA

Carole Holmberg, BS
Planetarium Director
Calusa Nature Center and Planetarium, Inc.
Fort Myers, FL

Tina C. Hopper
Science Writer
Rockwall, TX

Jonathan D. W. Kahl, PhD
Professor of Atmospheric Science
University of Wisconsin-Milwaukee
Milwaukee, WI

Nanette Kalis
Science Writer
Athens, OH

S. Page Keeley, MEd
Maine Mathematics and Science Alliance
Augusta, ME

Cindy Klevickis, PhD
Professor of Integrated Science and Technology
James Madison University
Harrisonburg, VA

Kimberly Fekany Lee, PhD
Science Writer
La Grange, IL

Michael Manga, PhD
Professor
University of California, Berkeley
Berkeley, CA

Devi Ried Mathieu
Science Writer
Sebastopol, CA

Elizabeth A. Nagy-Shadman, PhD
Geology Professor
Pasadena City College
Pasadena, CA

William D. Rogers, DA
Professor of Biology
Ball State University
Muncie, IN

Donna L. Ross, PhD
Associate Professor
San Diego State University
San Diego, CA

Marion B. Sewer, PhD
Assistant Professor
School of Biology
Georgia Institute of Technology
Atlanta, GA

Julia Meyer Sheets, PhD
Lecturer
School of Earth Sciences
The Ohio State University
Columbus, OH

Michael J. Singer, PhD
Professor of Soil Science
Department of Land, Air and Water Resources
University of California
Davis, CA

Karen S. Sottosanti, MA
Science Writer
Pickerington, Ohio

Paul K. Strode, PhD
I.B. Biology Teacher
Fairview High School
Boulder, CO

Jan M. Vermilye, PhD
Research Geologist
Seismo-Tectonic Reservoir Monitoring (STRM)
Boulder, CO

Judith A. Yero, MA
Director
Teacher's Mind Resources
Hamilton, MT

Dinah Zike, MEd
Author, Consultant,
Inventor of Foldables
Dinah Zike Academy;
Dinah-Might Adventures, LP
San Antonio, TX

Margaret Zorn, MS
Science Writer
Yorktown, VA

Consulting Authors

Alton L. Biggs
Biggs Educational Consulting
Commerce, TX

Ralph M. Feather, Jr., PhD
Assistant Professor
Department of Educational
Studies and Secondary
Education
Bloomsburg University
Bloomsburg, PA

Douglas Fisher, PhD
Professor of Teacher Education
San Diego State University
San Diego, CA

Edward P. Ortleb
Science/Safety Consultant
St. Louis, MO

Series Consultants

Science

Solomon Bililign, PhD
Professor
Department of Physics
North Carolina Agricultural
and Technical State University
Greensboro, NC

John Choinski
Professor
Department of Biology
University of Central Arkansas
Conway, AR

Anastasia Chopelas, PhD
Research Professor
Department of Earth and
Space Sciences
UCLA
Los Angeles, CA

David T. Crowther, PhD
Professor of Science Education
University of Nevada, Reno
Reno, NV

A. John Gatz
Professor of Zoology
Ohio Wesleyan University
Delaware, OH

Sarah Gille, PhD
Professor
University of California
San Diego
La Jolla, CA

David G. Haase, PhD
Professor of Physics
North Carolina State
University
Raleigh, NC

Janet S. Herman, PhD
Professor
Department of Environmental
Sciences
University of Virginia
Charlottesville, VA

David T. Ho, PhD
Associate Professor
Department of Oceanography
University of Hawaii
Honolulu, HI

Ruth Howes, PhD
Professor of Physics
Marquette University
Milwaukee, WI

Jose Miguel Hurtado, Jr., PhD
Associate Professor
Department of Geological
Sciences
University of Texas at El Paso
El Paso, TX

Monika Kress, PhD
Assistant Professor
San Jose State University
San Jose, CA

Mark E. Lee, PhD
Associate Chair & Assistant
Professor
Department of Biology
Spelman College
Atlanta, GA

Linda Lundgren
Science writer
Lakewood, CO

Series Consultants, continued

Keith O. Mann, PhD
Ohio Wesleyan University
Delaware, OH

Charles W. McLaughlin, PhD
Adjunct Professor of Chemistry
Montana State University
Bozeman, MT

Katharina Pahnke, PhD
Research Professor
Department of Geology and Geophysics
University of Hawaii
Honolulu, HI

Jesús Pando, PhD
Associate Professor
DePaul University
Chicago, IL

Hay-Oak Park, PhD
Associate Professor
Department of Molecular Genetics
Ohio State University
Columbus, OH

David A. Rubin, PhD
Associate Professor of Physiology
School of Biological Sciences
Illinois State University
Normal, IL

Toni D. Sauncy
Assistant Professor of Physics
Department of Physics
Angelo State University
San Angelo, TX

Malathi Srivatsan, PhD
Associate Professor of Neurobiology
College of Sciences and Mathematics
Arkansas State University
Jonesboro, AR

Cheryl Wistrom, PhD
Associate Professor of Chemistry
Saint Joseph's College
Rensselaer, IN

Reading

ReLeah Cossett Lent
Author/Educational Consultant
Blue Ridge, GA

Math

Vik Hovsepian
Professor of Mathematics
Rio Hondo College
Whittier, CA

Series Reviewers

Thad Boggs
Mandarin High School
Jacksonville, FL

Catherine Butcher
Webster Junior High School
Minden, LA

Erin Darichuk
West Frederick Middle School
Frederick, MD

Joanne Hedrick Davis
Murphy High School
Murphy, NC

Anthony J. DiSipio, Jr.
Octorara Middle School
Atglen, PA

Adrienne Elder
Tulsa Public Schools
Tulsa, OK

Series Reviewers, continued

Carolyn Elliott
Iredell-Statesville Schools
Statesville, NC

Christine M. Jacobs
Ranger Middle School
Murphy, NC

Jason O. L. Johnson
Thurmont Middle School
Thurmont, MD

Felecia Joiner
Stony Point Ninth Grade Center
Round Rock, TX

Joseph L. Kowalski, MS
Lamar Academy
McAllen, TX

Brian McClain
Amos P. Godby High School
Tallahassee, FL

Von W. Mosser
Thurmont Middle School
Thurmont, MD

Ashlea Peterson
Heritage Intermediate Grade Center
Coweta, OK

Nicole Lenihan Rhoades
Walkersville Middle School
Walkersvillle, MD

Maria A. Rozenberg
Indian Ridge Middle School
Davie, FL

Barb Seymour
Westridge Middle School
Overland Park, KS

Ginger Shirley
Our Lady of Providence Junior-Senior High School
Clarksville, IN

Curtis Smith
Elmwood Middle School
Rogers, AR

Sheila Smith
Jackson Public School
Jackson, MS

Sabra Soileau
Moss Bluff Middle School
Lake Charles, LA

Tony Spoores
Switzerland County Middle School
Vevay, IN

Nancy A. Stearns
Switzerland County Middle School
Vevay, IN

Kari Vogel
Princeton Middle School
Princeton, MN

Alison Welch
Wm. D. Slider Middle School
El Paso, TX

Linda Workman
Parkway Northeast Middle School
Creve Coeur, MO

Teacher Advisory Board

The Teacher Advisory Board gave the authors, editorial staff, and design team feedback on the content and design of the Student Edition. They provided valuable input in the development of *Glencoe ⓘScience*.

Frances J. Baldridge
Department Chair
Ferguson Middle School
Beavercreek, OH

Jane E. M. Buckingham
Teacher
Crispus Attucks Medical
Magnet High School
Indianapolis, IN

Elizabeth Falls
Teacher
Blalack Middle School
Carrollton, TX

Nelson Farrier
Teacher
Hamlin Middle School
Springfield, OR

Michelle R. Foster
Department Chair
Wayland Union
Middle School
Wayland, MI

Rebecca Goodell
Teacher
Reedy Creek Middle School
Cary, NC

Mary Gromko
Science Supervisor K–12
Colorado Springs District 11
Colorado Springs, CO

Randy Mousley
Department Chair
Dean Ray Stucky
Middle School
Wichita, KS

David Rodriguez
Teacher
Swift Creek Middle School
Tallahassee, FL

Derek Shook
Teacher
Floyd Middle Magnet School
Montgomery, AL

Karen Stratton
Science Coordinator
Lexington School District One
Lexington, SC

Stephanie Wood
Science Curriculum Specialist,
K–12
Granite School District
Salt Lake City, UT

Online Guide

Get ConnectED
connectED.mcgraw-hill.com

ConnectED
▶ **Your Digital Science Portal**

  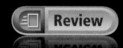  

Video	Audio	Review	Inquiry	WebQuest
See the science in real life through these exciting videos.	Click the link and you can listen to the text while you follow along.	Try these interactive tools to help you review the lesson concepts.	Explore concepts through hands-on and virtual labs.	These web-based challenges relate the concepts you're learning about to the latest news and research.

Digital and Print Solutions

The icons in your online student edition link you to interactive learning opportunities. Browse your online student book to find more.

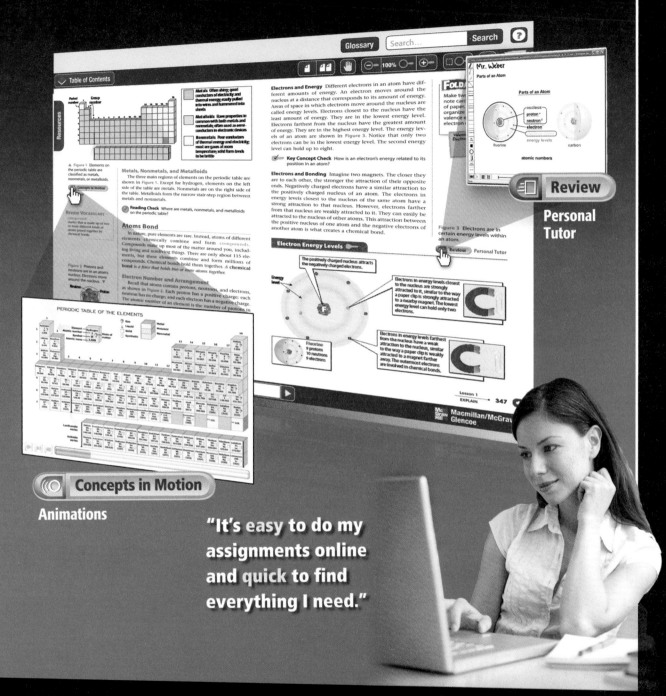

Personal Tutor

Animations

"It's easy to do my assignments online and quick to find everything I need."

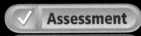

Check how well you understand the concepts with online quizzes and practice questions.

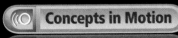

The textbook comes alive with animated explanations of important concepts.

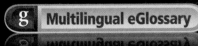

Read key vocabulary in 13 languages.

Treasure Hunt

Your science book has many features that will aid you in your learning. Some of these features are listed below. You can use the activity at the right to help you find these and other special features in the book.

- **THE BIG IDEA** can be found at the start of each chapter.
- The Reading Guide at the start of each lesson lists 🔑 **Key Concepts**, vocabulary terms, and online supplements to the content.
- **Connect ED** icons direct you to online resources such as animations, personal tutors, math practices, and quizzes.
- **Inquiry** Labs and Skill Practices are in each chapter.
- Your **FOLDABLES** help organize your notes.

START

1. What four margin items can help you build your vocabulary?

2. On what page does the glossary begin? What glossary is online?

3. In which Student Resource at the back of your book can you find a listing of Laboratory Safety Symbols?

4. Suppose you want to find a list of all the Launch Labs, MiniLabs, Skill Practices, and Labs, where do you look?

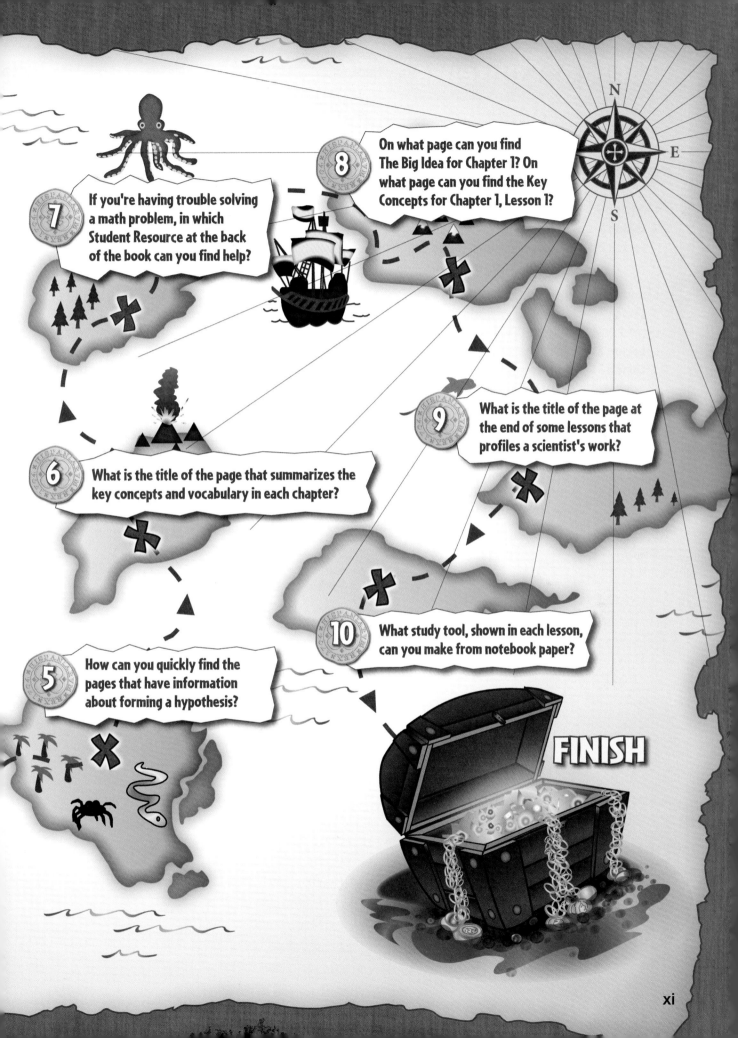

Table of Contents

Unit 3 **Properties of Matter** 306

Chapter 9 **Understanding the Atom** 310
Lesson 1 Discovering Parts of the Atom 312
Lesson 2 Protons, Neutrons, and Electrons—How Atoms Differ 325
 Lab Communicate Your Knowledge About the Atom 334

Chapter 10 **The Periodic Table** 342
Lesson 1 Using the Periodic Table 344
 Skill Practice How is the periodic table arranged? 353
Lesson 2 Metals 354
Lesson 3 Nonmetals and Metalloids 362
 Lab Alien Insect Periodic Table 370

Chapter 11 **Elements and Chemical Bonds** 378
Lesson 1 Electrons and Energy Levels 380
Lesson 2 Compounds, Chemical Formulas, and Covalent Bonds 389
 Skill Practice How can you model compounds? 396
Lesson 3 Ionic and Metallic Bonds 397
 Lab Ions in Solution 404

Table of Contents

Student Resources

Science Skill Handbook **SR-2**
   Scientific Methods SR-2
   Safety Symbols SR-11
   Safety in the Science Laboratory SR-12

Math Skill Handbook **SR-14**
   Math Review SR-14
   Science Application SR-24

Foldables Handbook **SR-29**

Reference Handbook **SR-40**
   Periodic Table of the Elements SR-40

Glossary **G-2**

Index **I-2**

Credits **C-2**

Inquiry

Launch Labs

9-1	What's in there?	313
9-2	How many different things can you make?	326
10-1	How can objects be organized?	345
10-2	What properties make metals useful?	355
10-3	What are some properties of nonmetals?	363
11-1	How is the periodic table organized?	381
11-2	How is a compound different from its elements?	390
11-3	How can atoms form compounds by gaining and losing electrons?	398

MiniLabs

9-1	How can you gather information about what you can't see?	320
9-2	How many penny isotopes do you have?	329
10-1	How does atom size change across a period?	351
10-2	How well do materials conduct thermal energy?	359
10-3	Which insulates better?	368
11-1	How does an electron's energy relate to its position in an atom?	386
11-2	How do compounds form?	394
11-3	How many ionic compounds can you make?	401

Inquiry

Inquiry Skill Practice

10-1 How is the periodic table arranged?.. 353
11-2 How can you model compounds?... 396

Inquiry Labs

9-2 Communicate Your Knowledge about the Atom.. 334
10-3 Alien Insect Periodic Table 370
11-3 Ions in Solution.. 404

Features

GREEN SCIENCE

11-1 New Green Airships 388

SCIENCE & SOCIETY

9-1 Subatomic Particles 324
10-2 Fireworks 361

Unit 3
Properties of Matter

"Sent to her room, Molly Cool dreams of escaping."

"If only she could change state and become a liquid, she could flow under the bedroom door and down the stairs..."

"...then flow to the fireplace where the heat would turn her into vapor and she could escape up the chimney."

"I'm free!"

"Hello birds!"

350 B.C.
Greek philosopher Aristotle defines an element as "one of those bodies into which other bodies can decompose, and that itself is not capable of being divided into another."

1704
Isaac Newton proposes that atoms attach to each other by some type of force.

1869
Dmitri Mendeleev publishes the first version of the periodic table.

1874
G. Johnstone Stoney proposes the existence of the electron, a subatomic particle that carries a negative electric charge, after experiments in electrochemistry.

1897
J.J. Thompson demonstrates the existence of the electron, proving Stoney's claim.

306 • Unit 3

1907
Physicists Hans Geiger and Ernest Marsden, under the direction of Ernest Rutherford, conduct the famous gold foil experiment. Rutherford concludes that the atom is mostly empty space and that most of the mass is concentrated in the atomic nucleus.

1918
Ernest Rutherford reports that the hydrogen nucleus has a positive charge, and he names it the proton.

1932
James Chadwick discovers the neutron, a subatomic particle with no electric charge and a mass slightly larger than a proton.

Inquiry
Visit ConnectED for this unit's STEM activity.

Unit 3 Nature of SCIENCE

Patterns

It's a bird! It's a plane! No, it's Venus! Besides the Sun, Venus is brighter than any other star or planet in the sky. It is often seen from Earth without the aid of a telescope, as shown in **Figure 1**. At certain times of the year, Venus can be seen in the early evening. At other times of the year, Venus is best seen in the morning or even during daylight hours.

Astronomers study the patterns of each planet's orbit and rotation. A pattern is a consistent plan or model used as a guide for understanding and predicting things. Studying the orbital patterns of planets allows scientists to predict the future position of each planet. By studying the pattern of Venus's orbit, astronomers can predict when Venus will be most visible from Earth. Astronomers also can predict when Venus will travel between Earth and the Sun, and be visible from Earth, as shown in **Figure 2**. This event is so rare that it has only occurred seven or eight times since the mid-1600s. Using patterns, scientists are able to predict the date when you will be able to see this event in the future.

▲ **Figure 1** Venus is often so bright in the morning sky that it has been nicknamed the morning star.

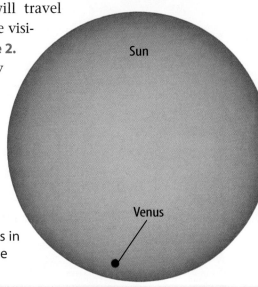

Figure 2 On June 8, 2004, observers around the world watched Venus pass in front of the Sun. This was the first time this event took place since 1882. ▶

Types of Patterns

Physical Patterns

A pattern that you can see and touch is a physical pattern. The crystalline structures of minerals are examples of physical patterns. When atoms form crystals, they produce structural, or physical, patterns. The crystal structure of the Star of India sapphire creates a pattern that reflects light in a stunning star shape.

Cyclic Patterns

An event that repeats many times again in a predictable order has a cyclic pattern. Since Earth's axis is tilted, the angle of the Sun's rays on your location on Earth changes as Earth orbits the Sun. This causes the seasons— winter, spring, summer, and fall— to occur in the same pattern every year.

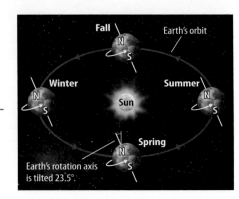

308 • Nature of Science

Patterns in Engineering

Engineers study patterns for many reasons, including to understand the physical properties of materials or to optimize the performance of their designs. Have you ever seen bricks with a pattern of holes through them? Clay bricks used in construction are fired, or baked, to make them stronger. Ceramic engineers understand that a regular pattern of holes in a brick assures that the brick is evenly fired and will not easily break.

Maybe you have seen a bridge constructed with a repeating pattern of large, steel triangles. Civil engineers, who design roads and bridges, know that the triangle is one of the strongest shapes in geometry. Engineers often use patterns of triangles in the structure of bridges to make them withstand heavy traffic and high winds.

Patterns in Physical Science

Scientists use patterns to explain past events or predict future events. At one time, only a few chemical elements were known. Chemists arranged the information they knew about these elements in a pattern according to the elements' properties. Scientists predicted the atomic numbers and the properties of elements that had yet to be discovered. These predictions made the discovery of new elements easier because scientists knew what properties to look for.

Look around. There are patterns everywhere—in art and nature, in the motion of the universe, in vehicles traveling on the roads, and in the processes of plant and animal growth. Analyzing patterns helps to understand the universe.

Patterns in Graphs
Scientists often graph their data to help identify patterns. For example, scientists might plot data from experiments on parachute nylon in graphs, such as the one below. Analyzing patterns on graphs then gives engineers information about how to design the strongest parachutes.

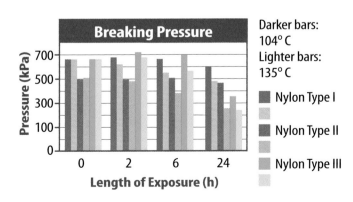

 **MiniLab**
15 minutes

How strong is your parachute?
Suppose you need to design a parachute. The graph to the left shows data for three types of parachute nylon. Each was tested to see how it weakens when exposed to different temperatures for different lengths of time. How would you use the patterns in the graph to design your parachute?

1. Write down the different experiments performed and how the variables changed in your Science Journal.

2. Write down all the patterns that you notice in the graph.

Analyze and Conclude

1. **Compare** Which nylon is weakest? What pattern helps you make this comparison?

2. **Identify** Which nylon is most affected by length of exposure to heat? What is its pattern on the graph?

3. **Select** Which nylon would you choose for your parachute? What pattern helped you make your decision?

Chapter 9

Understanding the Atom

THE BIG IDEA What are atoms, and what are they made of?

Inquiry All This to Study Tiny Particles?

This huge machine is called the Large Hadron Collider (LHC). It's like a circular racetrack for particles and is about 27 km long. The LHC accelerates particle beams to high speeds and then smashes them into each other. The longer the tunnel, the faster the beams move and the harder they smash together. Scientists study the tiny particles produced in the crash.

- How might scientists have studied matter before colliders were invented?
- What do you think are the smallest parts of matter?
- What are atoms, and what are they made of?

Get Ready to Read

What do you think?
Before you read, decide if you agree or disagree with each of these statements. As you read this chapter, see if you change your mind about any of the statements.

1. The earliest model of an atom contained only protons and electrons.
2. Air fills most of an atom.
3. In the present-day model of the atom, the nucleus of the atom is at the center of an electron cloud.
4. All atoms of the same element have the same number of protons.
5. Atoms of one element cannot be changed into atoms of another element.
6. Ions form when atoms lose or gain electrons.

ConnectED Your one-stop online resource

connectED.mcgraw-hill.com

- Video
- Audio
- Review
- Inquiry
- WebQuest
- Assessment
- Concepts in Motion
- Multilingual eGlossary

Lesson 1

Reading Guide

Key Concepts
ESSENTIAL QUESTIONS

- What is an atom?
- How would you describe the size of an atom?
- How has the atomic model changed over time?

Vocabulary

atom p. 315
electron p. 317
nucleus p. 320
proton p. 320
neutron p. 321
electron cloud p. 322

 Multilingual eGlossary

 Video   BrainPOP®

Discovering Parts of an Atom

Inquiry A Microscopic Mountain Range?

This photo shows a glimpse of the tiny particles that make up matter. A special microscope, invented in 1981, made this image. However, scientists knew these tiny particles existed long before they were able to see them. What are these tiny particles? How small do you think they are? How might scientists have learned so much about them before being able to see them?

Inquiry Launch Lab

10 minutes

What's in there?

When you look at a sandy beach from far away, it looks like a solid surface. You can't see the individual grains of sand. What would you see if you zoomed in on one grain of sand?

1. Read and complete a lab safety form.
2. Have your partner hold a **test tube** of **a substance,** filled to a height of 2–3 cm.
3. Observe the test tube from a distance of at least 2 m. Write a description of what you see in your Science Journal.
4. Pour about 1 cm of the substance onto a piece of **waxed paper.** Record your observations.
5. Use a **toothpick** to separate out one particle of the substance. Suppose you could zoom in. What do you think you would see? Record your ideas in your Science Journal.

Think About This

1. Do you think one particle of the substance is made of smaller particles? Why or why not?
2. **Key Concept** Do you think you could use a microscope to see what the particles are made of? Why or why not?

Early Ideas About Matter

Look at your hands. What are they made of? You might answer that your hands are made of things such as skin, bone, muscle, and blood. You might recall that each of these is made of even smaller structures called cells. Are cells made of even smaller parts? Imagine dividing something into smaller and smaller parts. What would you end up with?

Greek philosophers discussed and debated questions such as these more than 2,000 years ago. At the time, many thought that all matter is made of only four elements—fire, water, air, and earth, as shown in **Figure 1.** However, they weren't able to test their ideas because scientific tools and methods, such as experimentation, did not exist yet. The ideas proposed by the most influential philosophers usually were accepted over the ideas of less influential philosophers. One philosopher, Democritus (460–370 B.C.), challenged the popular idea of matter.

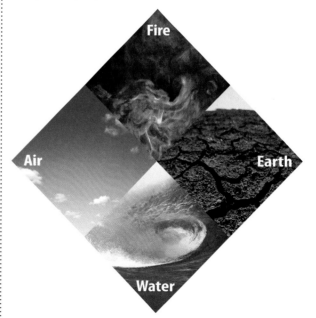

Figure 1 Most Greek philosophers believed that all matter is made of only four elements—fire, water, air, and earth.

Democritus

Democritus believed that matter is made of small, solid objects that cannot be divided, created, or destroyed. He called these objects *atomos*, from which the English word *atom* is derived. Democritus proposed that different types of matter are made from different types of atoms. For example, he said that smooth matter is made of smooth atoms. He also proposed that nothing is between these atoms except empty space. **Table 1** summarizes Democritus's ideas.

Although Democritus had no way to test his ideas, many of his ideas are similar to the way scientists describe the atom today. Because Democritus's ideas did not conform to the popular opinion and because they could not be tested scientifically, they were open for debate. One philosopher who challenged Democritus's ideas was Aristotle.

 Reading Check According to Democritus, what might atoms of gold look like?

Aristotle

Aristotle (384–322 B.C.) did not believe that empty space exists. Instead, he favored the more popular idea—that all matter is made of fire, water, air, and earth. Because Aristotle was so influential, his ideas were accepted. Democritus's ideas about atoms were not studied again for more than 2,000 years.

Dalton's Atomic Model

In the late 1700s, English schoolteacher and scientist John Dalton (1766–1844) revisited the idea of atoms. Since Democritus's time, advancements had been made in technology and scientific methods. Dalton made careful observations and measurements of chemical reactions. He combined data from his own scientific research with data from the research of other scientists to propose the atomic theory. **Table 1** lists ways that Dalton's atomic theory supported some of the ideas of Democritus.

Table 1 Similarities Between Democritus's and Dalton's Ideas

Democritus
1. Atoms are small solid objects that cannot be divided, created, or destroyed.
2. Atoms are constantly moving in empty space.
3. Different types of matter are made of different types of atoms.
4. The properties of the atoms determine the properties of matter.

John Dalton
1. All matter is made of atoms that cannot be divided, created, or destroyed.
2. During a chemical reaction, atoms of one element cannot be converted into atoms of another element.
3. Atoms of one element are identical to each other but different from atoms of another element.
4. Atoms combine in specific ratios.

▲ **Figure 2** If you could keep dividing a piece of aluminum, you eventually would have the smallest possible piece of aluminum—an aluminum atom.

The Atom

Today, scientists agree that matter is made of atoms with empty space between and within them. What is an atom? Imagine dividing the piece of aluminum shown in **Figure 2** into smaller and smaller pieces. At first you would be able to cut the pieces with scissors. But eventually you would have a piece that is too small to see—much smaller than the smallest piece you could cut with scissors. This small piece is an aluminum atom. An aluminum atom cannot be divided into smaller aluminum pieces. *An **atom** is the smallest piece of an element that still represents that element.*

 Key Concept Check What is a copper atom?

The Size of Atoms

Just how small is an atom? Atoms of different elements are different sizes, but all are very, very small. You cannot see atoms with just your eyes or even with most microscopes. Atoms are so small that about 7.5 trillion carbon atoms could fit into the period at the end of this sentence.

 Key Concept Check How would you describe the size of an atom?

Seeing Atoms

Scientific experiments verified that matter is made of atoms long before scientists were able to see atoms. However, the 1981 invention of a high-powered microscope, called a scanning tunneling microscope (STM), enabled scientists to see individual atoms for the first time. **Figure 3** shows an STM image. An STM uses a tiny, metal tip to trace the surface of a piece of matter. The result is an image of atoms on the surface.

Even today, scientists still cannot see inside an atom. However, scientists have learned that atoms are not the smallest particles of matter. In fact, atoms are made of much smaller particles. What are these particles, and how did scientists discover them if they could not see them?

Figure 3 A scanning tunneling microscope created this image. The yellow sphere is a manganese atom on the surface of gallium arsenide. ▼

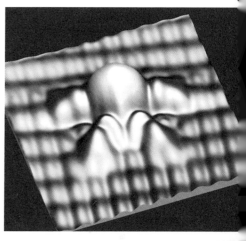

Thomson—Discovering Electrons

Not long after Dalton's findings, another English scientist, named J.J. Thomson (1856–1940), made some important discoveries. Thomson and other scientists of that time worked with cathode ray tubes. If you ever have seen a neon sign, an older computer monitor, or the color display on an ATM screen, you have seen a cathode ray tube. Thomson's cathode ray tube, shown in **Figure 4,** was a glass tube with pieces of metal, called electrodes, attached inside the tube. The electrodes were connected to wires, and the wires were connected to a battery. Thomson discovered that if most of the air was removed from the tube and electricity was passed through the wires, greenish-colored rays traveled from one electrode to the other end of the tube. What were these rays made of?

Negative Particles

Scientists called these rays cathode rays. Thomson wanted to know if these rays had an electric charge. To find out, he placed two plates on opposite sides of the tube. One plate was positively charged, and the other plate was negatively charged, as shown in **Figure 4.** Thomson discovered that these rays bent toward the positively charged plate and away from the negatively charged plate. Recall that opposite charges attract each other, and like charges repel each other. Thomson concluded that cathode rays are negatively charged.

Figure 4 As the cathode rays passed between the plates, they were bent toward the positive plate. Because opposite charges attract, the rays must be negatively charged.

 Reading Check If the rays were positively charged, what would Thomson have observed as they passed between the plates?

Thomson's Cathode Ray Tube Experiment

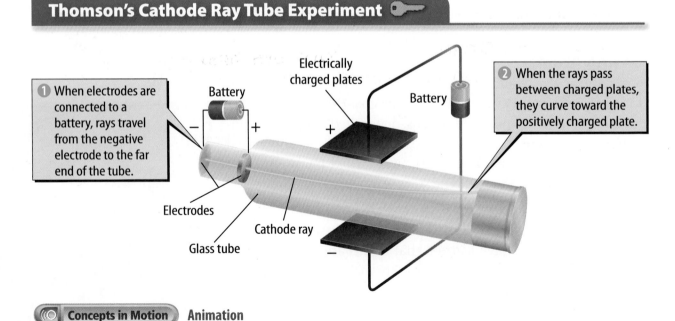

① When electrodes are connected to a battery, rays travel from the negative electrode to the far end of the tube.

② When the rays pass between charged plates, they curve toward the positively charged plate.

Concepts in Motion Animation

Parts of Atoms

Through more experiments, Thomson learned that these rays were made of particles that had mass. The mass of one of these particles was much smaller than the mass of the smallest atoms. This was surprising information to Thomson. Until then, scientists understood that the smallest particle of matter is an atom. But these rays were made of particles that were even smaller than atoms.

Where did these small, negatively charged particles come from? Thomson proposed that these particles came from the metal atoms in the electrode. Thomson discovered that identical rays were produced regardless of the kind of metal used to make the electrode. Putting these clues together, Thomson concluded that cathode rays were made of small, negatively charged particles. He called these particles electrons. *An* **electron** *is a particle with one negative charge (1−).* Because atoms are neutral, or not electrically charged, Thomson proposed that atoms also must contain a positive charge that balances the negatively charged electrons.

Thomson's Atomic Model

Thomson used this information to propose a new model of the atom. Instead of a solid, neutral sphere that was the same throughout, Thomson's model of the atom contained both positive and negative charges. He proposed that an atom was a sphere with a positive charge evenly spread throughout. Negatively charged electrons were mixed through the positive charge, similar to the way chocolate chips are mixed in cookie dough. **Figure 5** shows this model.

 Reading Check How did Thomson's atomic model differ from Dalton's atomic model?

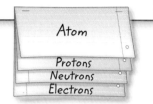

Use two sheets of paper to make a layered book. Label it as shown. Use it to organize your notes and diagrams on the parts of an atom.

WORD ORIGIN

electron
from Greek *electron*, means "amber," the physical force so called because it first was generated by rubbing amber. Amber is a fossilized substance produced by trees.

Thompson's Atomic Model

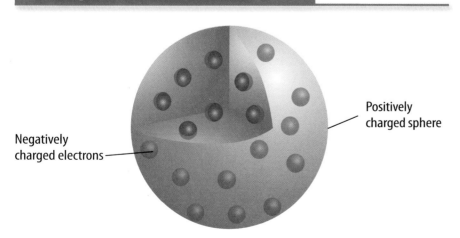

Figure 5 Thomson's model of the atom contained a positively charged sphere with negatively charged electrons within it.

Rutherford—Discovering the Nucleus

The discovery of electrons stunned scientists. Ernest Rutherford (1871–1937) was a student of Thomson's who eventually had students of his own. Rutherford's students set up experiments to test Thomson's atomic model and to learn more about what atoms contain. They discovered another surprise.

Rutherford's Predicted Result

Imagine throwing a baseball into a pile of table tennis balls. The baseball likely would knock the table tennis balls out of the way and continue moving in a relatively straight line. This is similar to what Rutherford's students expected to see when they shot alpha particles into atoms. Alpha particles are dense and positively charged. Because they are so dense, only another dense particle could deflect the path of an alpha particle. According to Thomson's model, the positive charge of the atom was too spread out and not dense enough to change the path of an alpha particle. Electrons wouldn't affect the path of an alpha particle because electrons didn't have enough mass. The result that Rutherford's students expected is shown in **Figure 6**.

 Reading Check Explain why Rutherford's students did not think an atom could change the path of an alpha particle.

Figure 6 The Thomson model of the atom did not contain a charge that was dense enough to change the path of an alpha particle. Rutherford expected the positive alpha particles to travel straight through the foil without changing direction.

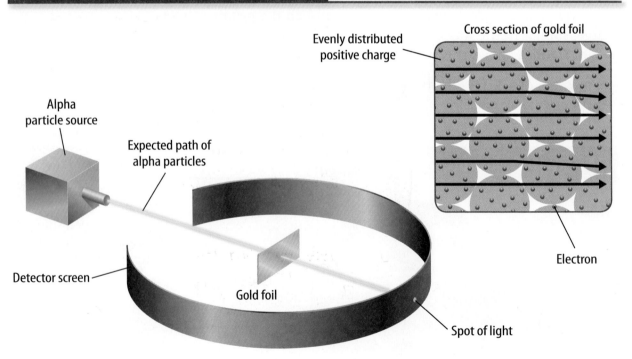

The Gold Foil Experiment

Rutherford's students went to work. They placed a source of alpha particles near a very thin piece of gold foil. Recall that all matter is made of atoms. Therefore, the gold foil was made of gold atoms. A screen surrounded the gold foil. When an alpha particle struck the screen, it created a spot of light. Rutherford's students could determine the path of the alpha particles by observing the spots of light on the screen.

The Surprising Result

Figure 7 shows what the students observed. Most of the particles did indeed travel through the foil in a straight path. However, a few particles struck the foil and bounced off to the side. And one particle in 10,000 bounced straight back! Rutherford later described this surprising result, saying it was almost as incredible as if you had fired a 38-cm shell at a piece of tissue paper and it came back and hit you. The alpha particles must have struck something dense and positively charged inside the nucleus. Thomson's model had to be refined.

 Key Concept Check Given the results of the gold foil experiment, how do you think an actual atom differs from Thomson's model?

Figure 7 Some alpha particles traveled in a straight path, as expected. But some changed direction, and some bounced straight back.

Visual Check What do the dots on the screen indicate?

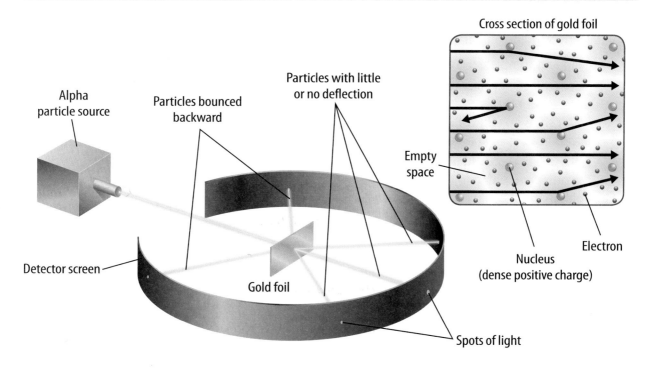

The Surprising Result

Rutherford's Atomic Model

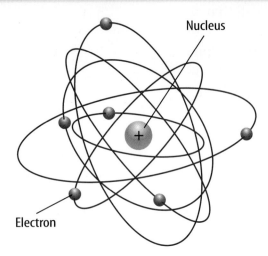

Figure 8 Rutherford's model contains a small, dense, positive nucleus. Tiny, negatively charged electrons travel in empty space around the nucleus.

Rutherford's Atomic Model

Because most alpha particles traveled through the foil in a straight path, Rutherford concluded that atoms are made mostly of empty space. The alpha particles that bounced backward must have hit a dense, positive mass. Rutherford concluded that *most of an atom's mass and positive charge is concentrated in a small area in the center of the atom called the* **nucleus**. Figure 8 shows Rutherford's atomic model. Additional research showed that the positive charge in the nucleus was made of positively charged particles called protons. *A* **proton** *is an atomic particle that has one positive charge (1+).* Negatively charged electrons move in the empty space surrounding the nucleus.

 Reading Check How did Rutherford explain the observation that some of the alpha particles bounced directly backward?

Inquiry MiniLab

20–30 minutes

How can you gather information about what you can't see?

Rutherford did his gold foil experiment to learn more about the structure of the atom. What can you learn by doing a similar investigation?

1. Read and complete a lab safety form.
2. Place a piece of white **newsprint** on a flat surface. Your teacher will place an upside-down **shoe box lid** with holes cut on opposite sides on the newsprint.
3. Place one end of a **ruler** on a **book,** with the other end pointing toward one of the holes in the shoe box lid. Roll a **marble** down the ruler and into one of the holes.
4. Team members should use **markers** to draw the path of the marble on the newsprint as it enters and leaves the lid. Predict the path of the marble under the lid. Draw it on the lid. Number the path *1*.
5. Take turns repeating steps 3 and 4 eight to ten times, numbering each path *2, 3, 4,* etc. Move the ruler and aim it in a slightly different direction each time.

Analyze and Conclude

1. **Recognize Cause and Effect** What caused the marble to change its path during some rolls and not during others?
2. **Draw Conclusions** How many objects are under the lid? Where are they located? Draw your answer.
3. **Key Concept** If the shoe box lid were an accurate model of the atom, what hypothesis would you make about the atom's structure?

Discovering Neutrons

The modern model of the atom was beginning to take shape. Rutherford's colleague, James Chadwick (1891–1974), also researched atoms and discovered that, in addition to protons, the nucleus also contained neutrons. A **neutron** is *a neutral particle that exists in the nucleus of an atom.*

Bohr's Atomic Model

Rutherford's model explained much of his students' experimental evidence. However, there were several observations that the model could not explain. For example, scientists noticed that if certain elements were heated in a flame, they gave off specific colors of light. Each color of light had a specific amount of energy. Where did this light come from? Niels Bohr (1885–1962), another student of Rutherford's, proposed an answer. Bohr studied hydrogen atoms because they contain only one electron. He experimented with adding electric energy to hydrogen and studying the energy that was released. His experiments led to a revised atomic model.

Electrons in the Bohr Model

Bohr's model is shown in **Figure 9**. Bohr proposed that electrons move in circular orbits, or energy levels, around the nucleus. Electrons in an energy level have a specific amount of energy. Electrons closer to the nucleus have less energy than electrons farther away from the nucleus. When energy is added to an atom, electrons gain energy and move from a lower energy level to a higher energy level. When the electrons return to the lower energy level, they release a specific amount of energy as light. This is the light that is seen when elements are heated.

Limitations of the Bohr Model

Bohr reasoned that if his model were accurate for atoms with one electron, it would be accurate for atoms with more than one electron. However, this was not the case. More research showed that, although electrons have specific amounts of energy, energy levels are not arranged in circular orbits. How do electrons move in an atom?

 Key Concept Check How did Bohr's atomic model differ from Rutherford's?

Figure 9 In Bohr's atomic model, electrons move in circular orbits around the atom. When an electron moves from a higher energy level to a lower energy level, energy is released—sometimes as light. Further research showed that electrons are not arranged in orbits.

Concepts in Motion Animation

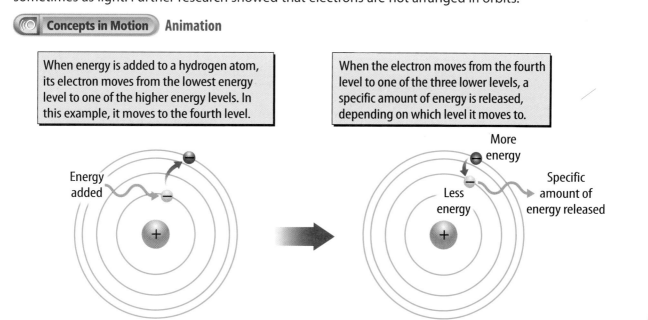

The Modern Atomic Model

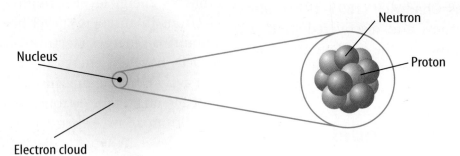

Figure 10 In this atom, electrons are more likely to be found closer to the nucleus than farther away.

Visual Check Why do you think this model of the atom doesn't show the electrons?

The Modern Atomic Model

In the modern atomic model, electrons form an electron cloud. *An **electron cloud** is an area around an atomic nucleus where an electron is most likely to be located.* Imagine taking a time-lapse photograph of bees around a hive. You might see a blurry cloud. The cloud might be denser near the hive than farther away because the bees spend more time near the hive.

In a similar way, electrons constantly move around the nucleus. It is impossible to know both the speed and exact location of an electron at a given moment in time. Instead, scientists only can predict the likelihood that an electron is in a particular location. The electron cloud shown in **Figure 10** is mostly empty space but represents the likelihood of finding an electron in a given area. The darker areas represent areas where electrons are more likely to be.

 Key Concept Check How has the model of the atom changed over time?

Quarks

You have read that atoms are made of smaller parts—protons, neutrons, and electrons. Are these particles made of even smaller parts? Scientists have discovered that electrons are not made of smaller parts. However, research has shown that protons and neutrons are made of smaller particles called quarks. Scientists theorize that there are six types of quarks. They have named these quarks up, down, charm, strange, top, and bottom. Protons are made of two up quarks and one down quark. Neutrons are made of two down quarks and one up quark. Just as the model of the atom has changed over time, the current model might also change with the invention of new technology that aids the discovery of new information.

Lesson 1 Review

 Assessment   Online Quiz

Visual Summary

If you were to divide an element into smaller and smaller pieces, the smallest piece would be an atom.

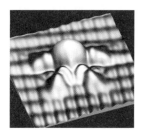

Atoms are so small that they can be seen only by using very powerful microscopes.

Scientists now know that atoms contain a dense, positive nucleus surrounded by an electron cloud.

FOLDABLES

Use your lesson Foldable to review the lesson. Save your Foldable for the project at the end of the chapter.

What do you think NOW?

You first read the statements below at the beginning of the chapter.

1. The earliest model of an atom contained only protons and electrons.
2. Air fills most of an atom.
3. In the present-day model of the atom, the nucleus of the atom is at the center of an electron cloud.

Did you change your mind about whether you agree or disagree with the statements? Rewrite any false statements to make them true.

Use Vocabulary

1. The smallest piece of the element gold is a gold _____.

2. **Write** a sentence that describes the nucleus of an atom.

3. **Define** *electron cloud* in your own words.

Understand Key Concepts

4. What is an atom mostly made of?
   A. air
   B. empty space
   C. neutrons
   D. protons

5. Why have scientists only recently been able to see atoms?
   A. Atoms are too small to see with ordinary microscopes.
   B. Early experiments disproved the idea of atoms.
   C. Scientists didn't know atoms existed.
   D. Scientists were not looking for atoms.

6. **Draw** Thomson's model of the atom, and label the parts of the drawing.

7. **Explain** how Rutherford's students knew that Thomson's model of the atom needed to change.

Interpret Graphics

8. **Contrast** Copy the graphic organizer below and use it to contrast the locations of electrons in Thomson's, Rutherford's, Bohr's, and the modern models of the atom.

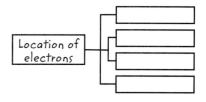

Critical Thinking

9. **Explain** what might have happened in Rutherford's experiment if he had used a thin sheet of copper instead of a thin sheet of gold.

Lesson 1
EVALUATE
323

SCIENCE & SOCIETY

Subatomic Particles

Welcome To The Particle Zoo

QUARKS
BOSONS
LEPTONS

Much has changed since Democritus and Aristotle studied atoms.

When Democritus and Aristotle developed ideas about matter, they probably never imagined the kinds of research being performed today! From the discovery of electrons, protons, and neutrons to the exploration of quarks and other particles, the atomic model continues to change.

You've learned about quarks, which make up protons and neutrons. But quarks are not the only kind of particles! In fact, some scientists call the collection of particles that have been discovered the particle zoo, because different types of particles have unique characteristics, just like the different kinds of animals in a zoo.

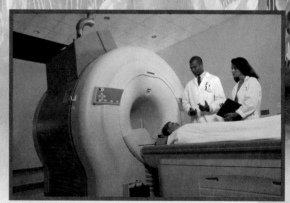

▲ MRIs are just one way in which particle physics technology is applied.

In addition to quarks, scientists have discovered a group of particles called leptons, which includes the electron. Gluons and photons are examples of bosons—particles that carry forces. Some particles, such as the Higgs Boson, have been predicted to exist but have yet to be observed in experiments.

Identifying and understanding the particles that make up matter is important work. However, it might be difficult to understand why time and money are spent to learn more about tiny subatomic particles. How can this research possibly affect everyday life? Research on subatomic particles has changed society in many ways. For example, magnetic resonance imaging (MRI), a tool used to diagnose medical problems, uses technology that was developed to study subatomic particles. Cancer treatments using protons, neutrons, and X-rays are all based on particle physics technology. And, in the 1990s, the need for particle physicists to share information with one another led to the development of the World Wide Web!

It's Your Turn

RESEARCH AND REPORT Learn more about research on subatomic particles. Find out about one recent discovery. Make a poster to share what you learn with your classmates.

Lesson 2

Protons, Neutrons, and Electrons—How Atoms Differ

Reading Guide

Key Concepts
ESSENTIAL QUESTIONS

- What happens during nuclear decay?
- How does a neutral atom change when its number of protons, electrons, or neutrons changes?

Vocabulary

atomic number p. 327
isotope p. 328
mass number p. 328
average atomic mass p. 329
radioactive p. 330
nuclear decay p. 331
ion p. 332

Multilingual eGlossary

Video BrainPOP®

Inquiry Is this glass glowing?

Under natural light, this glass vase is yellow. But when exposed to ultraviolet light, it glows green! That's because it is made of uranium glass, which contains small amounts of uranium, a radioactive element. Under ultraviolet light, the glass emits radiation.

Inquiry Launch Lab

15 minutes

How many different things can you make?

Many buildings are made of just a few basic building materials, such as wood, nails, and glass. You can combine those materials in many different ways to make buildings of various shapes and sizes. How many things can you make from three materials?

1. Read and complete a lab safety form.
2. Use **colored building blocks** to make as many different objects as you can with the following properties:
   - Each object must have a different number of red blocks.
   - Each object must have an equal number of red and blue blocks.
   - Each object must have at least as many yellow blocks as red blocks but can have no more than two extra yellow blocks.
3. As you complete each object, record in your Science Journal the number of each color of block used to make it. For example, R = 1; B = 1; Y = 2.
4. When time is called, compare your objects with others in the class.

Think About This

1. How many different objects did you make? How many different objects did the class make?
2. How many objects do you think you could make out of the three types of blocks?
3. **Key Concept** In what ways does changing the number of building blocks change the properties of the objects?

The Parts of the Atom

If you could see inside any atom, you probably would see the same thing—empty space surrounding a very tiny nucleus. A look inside the nucleus would reveal positively charged protons and neutral neutrons. Negatively charged electrons would be whizzing by in the empty space around the nucleus.

Table 2 compares the properties of protons, neutrons, and electrons. Protons and neutrons have about the same mass. The mass of electrons is much smaller than the mass of protons or neutrons. That means most of the mass of an atom is found in the nucleus. In this lesson, you will learn that, while all atoms contain protons, neutrons, and electrons, the numbers of these particles are different for different types of atoms.

Table 2 Properties of Protons, Neutrons, and Electrons

	Electron	Proton	Neutron
Symbol	e−	p	n
Charge	1−	1+	0
Location	electron cloud around the nucleus	nucleus	nucleus
Relative mass	1/1,840	1	1

Different Elements—Different Numbers of Protons

Look at the periodic table on the inside back cover of this book. Notice that there are more than 115 different elements. Recall that an element is a substance made from atoms that all have the same number of protons. For example, the element carbon is made from atoms that all have six protons. Likewise, all atoms that have six protons are carbon atoms. *The number of protons in an atom of an element is the element's* **atomic number.** The atomic number is the whole number listed with each element on the periodic table.

What makes an atom of one element different from an atom of another element? Atoms of different elements contain different numbers of protons. For example, oxygen atoms contain eight protons; nitrogen atoms contain seven protons. Different elements have different atomic numbers. **Figure 11** shows some common elements and their atomic numbers.

Neutral atoms of different elements also have different numbers of electrons. In a neutral atom, the number of electrons equals the number of protons. Therefore, the number of positive charges equals the number of negative charges.

Reading Check What two numbers can be used to identify an element?

FOLDABLES
Create a three-tab book and label it as shown. Use it to organize the three ways that atoms can differ.

Different Numbers of: Protons | Neutrons | Electrons

Figure 11 Atoms of different elements contain different numbers of protons.

Visual Check Explain the difference between an oxygen atom and a carbon atom.

Different Elements — Review | Personal Tutor

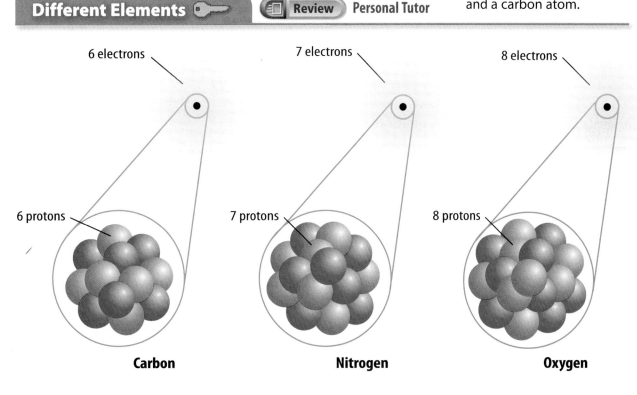

Carbon — 6 protons, 6 electrons
Nitrogen — 7 protons, 7 electrons
Oxygen — 8 protons, 8 electrons

Math Skills

Use Percentages

You can calculate the average atomic mass of an element if you know the percentage of each isotope in the element. Lithium (Li) contains 7.5% Li-6 and 92.5% Li-7. What is the average atomic mass of Li?

1. Divide each percentage by 100 to change to decimal form.
   $\frac{7.5\%}{100} = 0.075$
   $\frac{92.5\%}{100} = 0.925$

2. Multiply the mass of each isotope by its decimal percentage.
   $6 \times 0.075 = 0.45$
   $7 \times 0.925 = 6.475$

3. Add the values together to get the average atomic mass.
   $0.45 + 6.475 = 6.93$

Practice

Nitrogen (N) contains 99.63% N-14 and 0.37% N-15. What is the average atomic mass of nitrogen?

 Review
- Math Practice
- Personal Tutor

WORD ORIGIN

isotope
from Greek *isos*, means "equal"; and *topos*, means "place"

Table 3 Naturally Occurring Isotopes of Carbon

Isotope	Carbon-12 Nucleus	Carbon-13 Nucleus	Carbon-14 Nucleus
Abundance	98.89%	<1.11%	<0.01%
Protons	6	6	6
Neutrons	+6	+7	+8
Mass Number	12	13	14

Neutrons and Isotopes

You have read that atoms of the same element have the same numbers of protons. However, atoms of the same element can have different numbers of neutrons. For example, carbon atoms all have six protons, but some carbon atoms have six neutrons, some have seven neutrons, and some have eight neutrons. These three different types of carbon atoms, shown in **Table 3**, are called isotopes. **Isotopes** *are atoms of the same element that have different numbers of neutrons.* Most elements have several isotopes.

Protons, Neutrons, and Mass Number

The **mass number** *of an atom is the sum of the number of protons and neutrons in an atom.* This is shown in the following equation.

Mass number = number of protons + number of neutrons

Any one of these three quantities can be determined if you know the value of the other two quantities. For example, to determine the mass number of an atom, you must know the number of neutrons and the number of protons in the atom.

The mass numbers of the isotopes of carbon are shown in **Table 3**. An isotope often is written with the element name followed by the mass number. Using this method, the isotopes of carbon are written carbon-12, carbon-13, and carbon-14.

 Reading Check How do two different isotopes of the same element differ?

Average Atomic Mass

You might have noticed that the periodic table does not list mass numbers or the numbers of neutrons. This is because a given element can have several isotopes. However, you might notice that there is a decimal number listed with most elements, as shown in **Figure 12.** This decimal number is the average atomic mass of the element. The **average atomic mass** of an element is the average mass of the element's isotopes, weighted according to the abundance of each isotope.

Table 3 shows the three isotopes of carbon. The average atomic mass of carbon is 12.01. Why isn't the average atomic mass 13? After all, the average of the mass numbers 12, 13, and 14 is 13. The average atomic mass is weighted based on each isotope's abundance—how much of each isotope is present on Earth. Almost 99 percent of Earth's carbon is carbon-12. That is why the average atomic mass is close to 12.

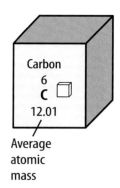

Figure 12 The element carbon has several isotopes. The decimal number 12.01 is the average atomic mass of these isotopes.

✓ **Reading Check** What does the term *weighted average* mean?

Inquiry MiniLab

20 minutes

How many penny isotopes do you have?

All pennies look similar, and all have a value of one cent. But do they have the same mass? Let's find out.

1. Read and complete a lab safety form.
2. Copy the data table into your Science Journal.

Penny Sample	Mass of 10 pennies (g)	Average mass of 1 penny (g)
Pre-1982		
Post-1982		
Unknown mix		

3. Use a **balance** to find the mass of **10 pennies minted before 1982.** Record the mass in the data table.
4. Divide the mass by 10 to find the average mass of one penny. Record the answer.
5. Repeat steps 3 and 4 with **10 pennies minted after 1982.**
6. Have a team member combine pre- and post-1982 pennies for a total of 10 pennies. Find the mass of the ten pennies and the average mass of one penny. Record your observations.

Analyze and Conclude

1. **Compare and Contrast** How did the average mass of pre- and post-1982 pennies compare?
2. **Draw Conclusions** How many pennies of each type were in the 10 pennies assembled by your partner? How do you know?
3. 🔑 **Key Concept** How does this activity relate to the way in which scientists calculate the average atomic mass of an element?

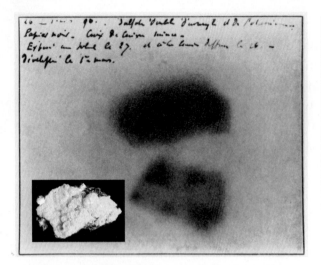

▲ **Figure 13** The black and white photo shows Henri Becquerel's photographic plate. The dark area on the plate was exposed to radiation given off by uranium in the mineral even though the mineral was not exposed to sunlight.

ACADEMIC VOCABULARY

spontaneous
(adjective) occurring without external force or cause

Figure 14 Marie Curie studied radioactivity and discovered two new radioactive elements—polonium and radium. ▼

Radioactivity

More than 1,000 years ago, people tried to change lead into gold by performing chemical reactions. However, none of their reactions were successful. Why not? Today, scientists know that a chemical reaction does not change the number of protons in an atom's nucleus. If the number of protons does not change, the element does not change. But in the late 1800s, scientists discovered that some elements change into other elements **spontaneously.** How does this happen?

An Accidental Discovery

In 1896, a scientist named Henri Becquerel (1852–1908) studied minerals containing the element uranium. When these minerals were exposed to sunlight, they gave off a type of energy that could pass through paper. If Becquerel covered a photographic plate with black paper, this energy would pass through the paper and expose the film. One day, Becquerel left the mineral next to a wrapped, unexposed plate in a drawer. Later, he opened the drawer, unwrapped the plate, and saw that the plate contained an image of the mineral, as shown in **Figure 13.** The mineral spontaneously emitted energy, even in the dark! Sunlight wasn't required. What was this energy?

Radioactivity

Becquerel shared his discovery with fellow scientists Pierre and Marie Curie. Marie Curie (1867–1934), shown in **Figure 14,** called *elements that spontaneously emit radiation* **radioactive.** Becquerel and the Curies discovered that the radiation released by uranium was made of energy and particles. This radiation came from the nuclei of the uranium atoms. When this happens, the number of protons in one atom of uranium changes. When uranium releases radiation, it changes to a different element!

Types of Decay

Radioactive elements contain unstable nuclei. **Nuclear decay** *is a process that occurs when an unstable atomic nucleus changes into another more stable nucleus by emitting radiation.* Nuclear decay can produce three different types of radiation—alpha particles, beta particles, and gamma rays. **Figure 15** compares the three types of nuclear decay.

Alpha Decay An alpha particle is made of two protons and two neutrons. When an atom releases an alpha particle, its atomic number decreases by two. Uranium-238 decays to thorium-234 through the process of alpha decay.

Beta Decay When beta decay occurs, a neutron in an atom changes into a proton and a high-energy electron called a beta particle. The new proton becomes part of the nucleus, and the beta particle is released. In beta decay, the atomic number of an atom increases by one because it has gained a proton.

Gamma Decay Gamma rays do not contain particles, but they do contain a lot of energy. In fact, gamma rays can pass through thin sheets of lead! Because gamma rays do not contain particles, the release of gamma rays does not change one element into another element.

 Key Concept Check What happens during radioactive decay?

Uses of Radioactive Isotopes

The energy released by radioactive decay can be both harmful and beneficial to humans. Too much radiation can damage or destroy living cells, making them unable to function properly. Some organisms contain cells, such as cancer cells, that are harmful to the organism. Radiation therapy can be beneficial to humans by destroying these harmful cells.

Figure 15 Alpha and beta decay change one element into another element.

Visual Check Explain the change in atomic number for each type of decay.

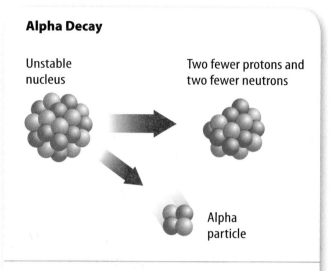

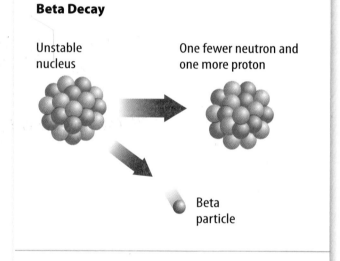

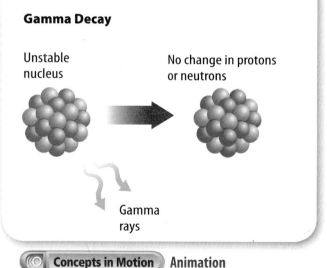

Ions—Gaining or Losing Electrons

What happens to a neutral atom if it gains or loses electrons? Recall that a neutral atom has no overall charge. This is because it contains equal numbers of positively charged protons and negatively charged electrons. When electrons are added to or removed from an atom, that atom becomes an ion. *An **ion** is an atom that is no longer neutral because it has gained or lost electrons.* An ion can be positively or negatively charged depending on whether it has lost or gained electrons.

Positive Ions

When a neutral atom loses one or more electrons, it has more protons than electrons. As a result, it has a positive charge. An atom with a positive charge is called a positive ion. A positive ion is represented by the element's symbol followed by a superscript plus sign (⁺). For example, **Figure 16** shows how sodium (Na) becomes a positive sodium ion (Na⁺).

Negative Ions

When a neutral atom gains one or more electrons, it now has more electrons than protons. As a result, the atom has a negative charge. An atom with a negative charge is called a negative ion. A negative ion is represented by the element's symbol followed by a superscript negative sign (⁻). **Figure 16** shows how fluorine (F) becomes a fluoride ion (F⁻).

 Key Concept Check How does a neutral atom change when its number of protons, electrons, or neutrons changes?

Figure 16 An ion is formed when a neutral atom gains or loses an electron.

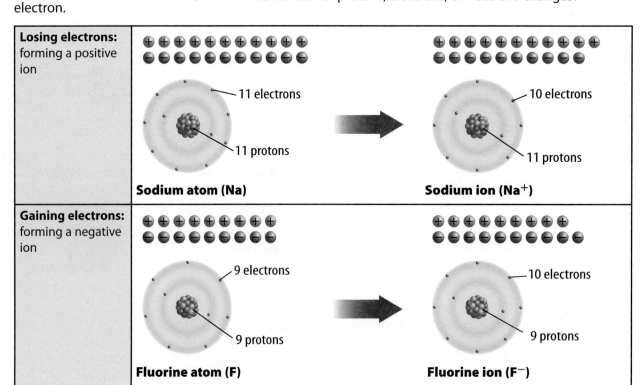

Lesson 2 Review

Assessment — Online Quiz
Inquiry — Virtual Lab

Visual Summary

Carbon — **Nitrogen**

Different elements contain different numbers of protons.

Isotopes

Two isotopes of a given element contain different numbers of neutrons.

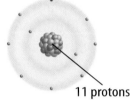

10 electrons

11 protons

Sodium ion (Na$^+$)

When a neutral atom gains or loses an electron, it becomes an ion.

FOLDABLES

Use your lesson Foldable to review the lesson. Save your Foldable for the project at the end of the chapter.

What do you think NOW?

You first read the statements below at the beginning of the chapter.

4. All atoms of the same element have the same number of protons.

5. Atoms of one element cannot be changed into atoms of another element.

6. Ions form when atoms lose or gain electrons.

Did you change your mind about whether you agree or disagree with the statements? Rewrite any false statements to make them true.

Use Vocabulary

1 The number of protons in an atom of an element is its _____.

2 Nuclear decay occurs when an unstable atomic nucleus changes into another nucleus by emitting _____.

3 Describe how two isotopes of nitrogen differ from two nitrogen ions.

Understand Key Concepts

4 An element's average atomic mass is calculated using the masses of its
A. electrons.   C. neutrons.
B. isotopes.   D. protons.

5 Compare and contrast oxygen-16 and oxygen-17.

6 Show what happens to the electrons of a neutral calcium atom (Ca) when it is changed into a calcium ion (Ca^{2+}).

Interpret Graphics

7 Contrast Copy and fill in this graphic organizer to contrast how different elements, isotopes, and ions are produced.

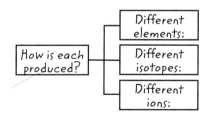

Critical Thinking

8 Consider Find two neighboring elements on the periodic table whose positions would be reversed if they were arranged by atomic mass instead of atomic number.

9 Infer Can an isotope also be an ion?

Math Skills — Review — Math Practice

10 A sample of copper (Cu) contains 69.17% Cu-63. The remaining copper atoms are Cu-65. What is the average atomic mass of copper?

Inquiry Lab

2 class periods

Communicate Your Knowledge About the Atom

Materials

computer

creative building materials

drawing and modeling materials

office supplies

Also needed:
recording devices, software, or other equipment for multimedia presentations

In this chapter, you have learned many things about atoms. Suppose that you are asked to take part in an atom fair. Each exhibit in the fair will help visitors understand something new about atoms in an exciting and interesting way. What will your exhibit be like? Will you hold a mock interview with Democritus or Rutherford? Will you model a famous experiment and explain its conclusion? Can visitors assemble or make models of their own atoms? Is there a multimedia presentation? Will your exhibit be aimed at children or adults? The choice is yours!

Question

Which concepts about the atom did you find most interesting? How can you present the information in exciting, creative, and perhaps unexpected ways? Think about whether you will present the information yourself or have visitors interact with the exhibit.

Procedure

1. In your Science Journal, write your ideas about the following questions:

- What specific concepts about the atom do you want your exhibit to teach?
- How will you present the information to your visitors?
- How will you make the information exciting and interesting to keep your visitors' attention?

2. Outline the steps in preparing for your exhibit.

- What materials and equipment will you need?
- How much time will it take to prepare each part of your exhibit?
- Will you involve anyone else? For example, if you are going to interview a scientist about an early model of the atom, who will play the scientist? What questions will you ask?

③ Have your teacher approve your plan.

④ Follow the steps you outlined, and prepare your exhibit.

⑤ Ask family members and/or several friends to view your exhibit. Invite them to tell you what they've learned from your exhibit. Compare this with what you had expected to teach in your exhibit.

⑥ Ask your friends for feedback about what could be more effective in teaching the concepts you intend to teach.

⑦ Modify your exhibit to make it more effective.

⑧ If you can, present your exhibit to visitors of various ages, including teachers and students from other classes. Have visitors fill out a comment form.

Lab Tips

☑ For very young visitors, you might draw a picture book about the three parts of the atom and make them into cartoon characters.

☑ Plan your exhibit so visitors spend about 5 minutes there. Don't try to present too much or too little information.

☑ Remember that a picture is worth a thousand words!

Analyze and Conclude

⑨ **Infer** What did visitors to your exhibit find the most and least interesting? How do you know?

⑩ **Predict** What would you do differently if you had a chance to plan your exhibit again? Why?

⑪ **The Big Idea** In what ways did your exhibit help visitors understand the current model of the atom?

Communicate Your Results

After the fair is over, discuss the visitors' comments and how you might improve organization or individual experiences if you were to do another fair.

Design and describe an interactive game or activity that would teach the same concept you used for your exhibit. Invite other students to comment on your design.

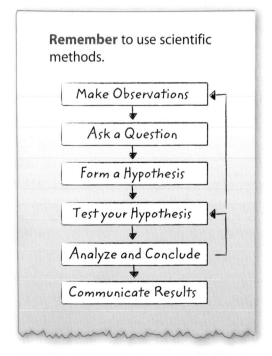

Remember to use scientific methods.

- Make Observations
- Ask a Question
- Form a Hypothesis
- Test your Hypothesis
- Analyze and Conclude
- Communicate Results

Chapter 9 Study Guide

An atom is the smallest unit of an element and is made mostly of empty space. It contains a tiny nucleus surrounded by an electron cloud.

Key Concepts Summary

Lesson 1: Discovering Parts of the Atom

- If you were to divide an element into smaller and smaller pieces, the smallest piece would be an **atom**.
- Atoms are so small that they can be seen only by powerful scanning microscopes.
- The first model of the atom was a solid sphere. Now, scientists know that an atom contains a dense positive **nucleus** surrounded by an **electron cloud**.

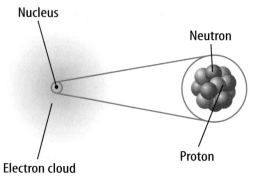

Vocabulary

atom p. 315
electron p. 317
nucleus p. 320
proton p. 320
neutron p. 321
electron cloud p. 322

Lesson 2: Protons, Neutrons, and Electrons—How Atoms Differ

- **Nuclear decay** occurs when an unstable atomic nucleus changes into another more stable nucleus by emitting radiation.
- Different elements contain different numbers of protons. Two **isotopes** of the same element contain different numbers of neutrons. When a neutral atom gains or loses an electron, it becomes an **ion**.

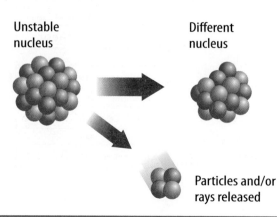

atomic number p. 327
isotope p. 328
mass number p. 328
average atomic mass p. 329
radioactive p. 330
nuclear decay p. 331
ion p. 332

Study Guide

- Personal Tutor
- Vocabulary eGames
- Vocabulary eFlashcards

FOLDABLES Chapter Project

Assemble your lesson Foldables as shown to make a Chapter Project. Use the project to review what you have learned in this chapter.

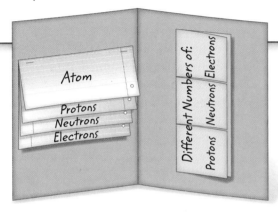

Use Vocabulary

1. A(n) _____ is a very small particle that is the basic unit of matter.

2. Electrons in an atom move throughout the _____ surrounding the nucleus.

3. _____ is the weighted average mass of all of an element's isotopes.

4. All atoms of a given element have the same number of _____.

5. When _____ occurs, one element is changed into another element.

6. Isotopes have the same _____, but different mass numbers.

Link Vocabulary and Key Concepts

 Interactive Concept Map

Copy this concept map, and then use vocabulary terms from the previous page to complete the concept map.

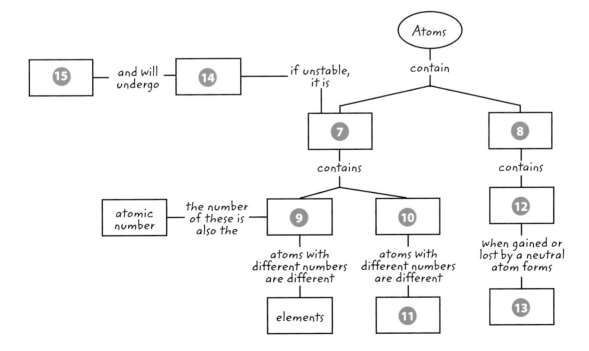

Chapter 9 Review

Understand Key Concepts

1. Which part of an atom makes up most of its volume?
   A. its electron cloud
   B. its neutrons
   C. its nucleus
   D. its protons

2. What did Democritus believe an atom was?
   A. a solid, indivisible object
   B. a tiny particle with a nucleus
   C. a nucleus surrounded by an electron cloud
   D. a tiny nucleus with electrons surrounding it

3. If an ion contains 10 electrons, 12 protons, and 13 neutrons, what is the ion's charge?
   A. 2−
   B. 1−
   C. 2+
   D. 3+

4. J.J. Thomson's experimental setup is shown below.

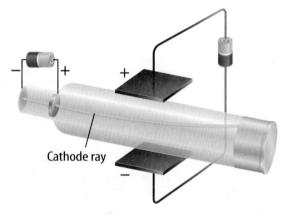

   What is happening to the cathode rays?
   A. They are attracted to the negative plate.
   B. They are attracted to the positive plate.
   C. They are stopped by the plates.
   D. They are unaffected by either plate.

5. How many neutrons does iron-59 have?
   A. 30
   B. 33
   C. 56
   D. 59

6. Why were Rutherford's students surprised by the results of the gold foil experiment?
   A. They didn't expect the alpha particles to bounce back from the foil.
   B. They didn't expect the alpha particles to continue in a straight path.
   C. They expected only a few alpha particles to bounce back from the foil.
   D. They expected the alpha particles to be deflected by electrons.

7. Which determines the identity of an element?
   A. its mass number
   B. the charge of the atom
   C. the number of its neutrons
   D. the number of its protons

8. The figure below shows which of the following?

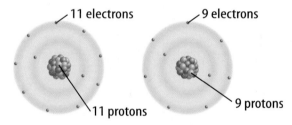

   A. two different elements
   B. two different ions
   C. two different isotopes
   D. two different protons

9. How is Bohr's atomic model different from Rutherford's model?
   A. Bohr's model has a nucleus.
   B. Bohr's model has electrons.
   C. Electrons in Bohr's model are located farther from the nucleus.
   D. Electrons in Bohr's model are located in circular energy levels.

Chapter Review

Assessment — Online Test Practice

Critical Thinking

10 Consider what would have happened in the gold foil experiment if Dalton's theory had been correct.

11 Contrast How does Bohr's model of the atom differ from the present-day atomic model?

12 Describe the electron cloud using your own analogy.

13 Summarize how radioactive decay can produce new elements.

14 Hypothesize What might happen if a negatively charged ion comes into contact with a positively charged ion?

15 Infer Why isn't mass number listed with each element on the periodic table?

16 Explain How is the average atomic mass calculated?

17 Infer Oxygen has three stable isotopes.

Isotope	Average Atomic Mass
Oxygen-16	0.99757
Oxygen-17	0.00038
Oxygen-18	0.00205

What can you determine about the average atomic mass of oxygen without calculating it?

Writing in Science

18 Write a newspaper article that describes how the changes in the atomic model provide an example of the scientific process in action.

REVIEW THE BIG IDEA

19 Describe the current model of the atom. Explain the size of atoms. Also explain the charge, the location, and the size of protons, neutrons, and electrons.

20 Summarize The Large Hadron Collider, shown below, is continuing the study of matter and energy. Use a set of four drawings to summarize how the model of the atom changed from Thomson, to Rutherford, to Bohr, to the modern model.

Math Skills

Review — Math Practice

Use Percentages

Use the information in the table to answer questions 21 and 22.

Magnesium (Mg) Isotope	Percent Found in Nature
Mg-24	78.9%
Mg-25	10.0%
Mg-26	

21 What is the percentage of Mg-26 found in nature?

22 What is the average atomic mass of magnesium?

Standardized Test Practice

Record your answers on the answer sheet provided by your teacher or on a sheet of paper.

Multiple Choice

1. Which best describes an atom?
   A. a particle with a single negative charge
   B. a particle with a single positive charge
   C. the smallest particle that still represents a compound
   D. the smallest particle that still represents an element

Use the figure below to answer questions 2 and 3.

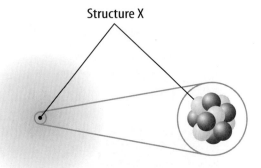

2. What is Structure X?
   A. an electron
   B. a neutron
   C. a nucleus
   D. a proton

3. Which best describes Structure X?
   A. most of the atom's mass, neutral charge
   B. most of the atom's mass, positive charge
   C. very small part of the atom's mass, negative charge
   D. very small part of the atom's mass, positive charge

4. Which is true about the size of an atom?
   A. It can only be seen using a scanning tunneling microscope.
   B. It is about the size of the period at the end of this sentence.
   C. It is large enough to be seen using a magnifying lens.
   D. It is too small to see with any type of microscope.

Use the figure below to answer question 5.

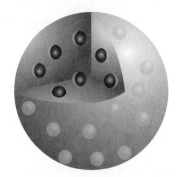

5. Whose model for the atom is shown?
   A. Bohr's
   B. Dalton's
   C. Rutherford's
   D. Thomson's

6. What structure did Rutherford discover?
   A. the atom
   B. the electron
   C. the neutron
   D. the nucleus

Standardized Test Practice

Use the table below to answer questions 7–9.

Particle	Number of Protons	Number of Neutrons	Number of Electrons
1	4	5	2
2	5	5	5
3	5	6	5
4	6	6	6

7 What is atomic number of particle 3?
- **A** 3
- **B** 5
- **C** 6
- **D** 11

8 Which particles are isotopes of the same element?
- **A** 1 and 2
- **B** 2 and 3
- **C** 2 and 4
- **D** 3 and 4

9 Which particle is an ion?
- **A** 1
- **B** 2
- **C** 3
- **D** 4

10 Which reaction starts with a neutron and results in the formation of a proton and a high-energy electron?
- **A** alpha decay
- **B** beta decay
- **C** the formation of positive ion
- **D** the formation of negative ion

Constructed Response

Use the figure below to answer questions 11 and 12.

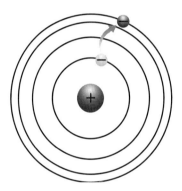

11 Identify the atomic model shown in the figure, and describe its characteristics.

12 How does this atomic model differ from the modern atomic model?

13 Compare two different neutral isotopes of the same element. Then compare two different ions of the same element. What do all of these particles have in common?

14 How does nuclear decay differ from the formation of ions? What parts of the atom are affected in each type of change?

NEED EXTRA HELP?														
If You Missed Question...	1	2	3	4	5	6	7	8	9	10	11	12	13	14
Go to Lesson...	1	1	1	1	1	1	2	2	2	2	1	1	2	2

Chapter 10

The Periodic Table

 How is the periodic table used to classify and provide information about all known elements?

Inquiry What makes this balloon so special?

Things are made out of specific materials for a reason. A weather balloon can rise high in the atmosphere and gather weather information. The plastic that forms this weather balloon and the helium gas that fills it were chosen after scientists researched and studied the properties of these materials.

- What property of helium do you think makes the balloon rise through the air?
- How do you think the periodic table is a useful tool when determining properties of different materials?

Get Ready to Read

What do you think?

Before you read, decide if you agree or disagree with each of these statements. As you read this chapter, see if you change your mind about any of the statements.

1. The elements on the periodic table are arranged in rows in the order they were discovered.
2. The properties of an element are related to the element's location on the periodic table.
3. Fewer than half of the elements are metals.
4. Metals are usually good conductors of electricity.
5. Most of the elements in living things are nonmetals.
6. Even though they look very different, oxygen and sulfur share some similar properties.

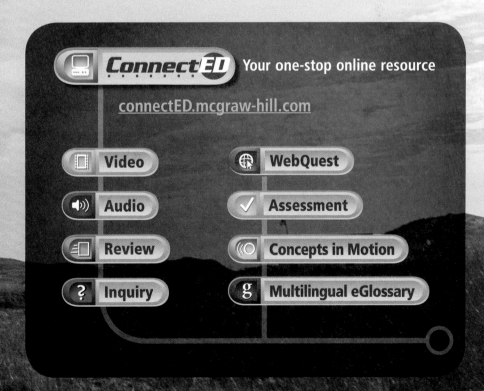

ConnectED — Your one-stop online resource

connectED.mcgraw-hill.com

- Video
- Audio
- Review
- Inquiry
- WebQuest
- Assessment
- Concepts in Motion
- Multilingual eGlossary

Lesson 1

Using the Periodic Table

Reading Guide

Key Concepts
ESSENTIAL QUESTIONS

- How are elements arranged on the periodic table?
- What can you learn about elements from the periodic table?

Vocabulary

periodic table p. 345
group p. 350
period p. 350

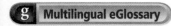 Multilingual eGlossary

 Video   BrainPOP®

Inquiry Same Information?

You probably have seen a copy of a table that is used to organize the elements. Does it look like this chart? There is no specific shape that a chart of elements must have. However, the relationships among the elements in the chart are important.

Launch Lab

15 minutes

How can objects be organized?

What would it be like to shop at a grocery store where all the products are mixed up on the shelves? Maybe cereal is next to the dish soap and bread is next to the canned tomatoes. It would take a long time to find the groceries that you needed. How does organizing objects help you to find and use what you need?

1. Read and complete a lab safety form.
2. Empty the **interlocking plastic bricks** from the **plastic bag** onto your desk and observe their properties. Think about ways you might group and sequence the bricks so that they are organized.
3. Organize the bricks according to your plan.
4. Compare your pattern of organization with those used by several other students.

Think About This

1. Describe in your Science Journal the way you grouped your bricks. Why did you choose that way of grouping?
2. Describe how you sequenced the bricks.
3. **Key Concept** How does organizing things help you to use them more easily?

What is the periodic table?

The "junk drawer" in **Figure 1** is full of pens, notepads, rubber bands, and other supplies. It would be difficult to find a particular item in this messy drawer. How might you organize it? First, you might dump the contents onto the counter. Then you could sort everything into piles. Pens and pencils might go into one pile. Notepads and paper go into another. Organizing the contents of the drawer makes it easier to find the things you need, also shown in **Figure 1**.

Just as sorting helped to organize the objects in the junk drawer, sorting can help scientists organize information about the elements. Recall that there are more than 100 elements, each with a unique set of physical and chemical properties.

Scientists use a table called the periodic (pihr ee AH dihk) table to organize elements. *The* **periodic table** *is a chart of the elements arranged into rows and columns according to their physical and chemical properties.* It can be used to determine the relationships among the elements.

In this chapter, you will read about how the periodic table was developed. You will also read about how you can use the periodic table to learn about the elements.

Figure 1 Sorting objects by their similarities makes it easier to find what you need.

Developing a Periodic Table

In 1869 a Russian chemist and teacher named Dimitri Mendeleev (duh MEE tree • men duh LAY uf) was working on a way to classify elements. At that time, more than 60 elements had been discovered. He studied the physical properties such as density, color, melting point, and atomic mass of each element. Mendeleev also noted chemical properties such as how each element reacted with other elements. Mendeleev arranged the elements in a list using their atomic masses. He noticed that the properties of the elements seemed to repeat in a pattern.

When Mendeleev placed his list of elements into a table, he arranged them in rows of increasing atomic mass. Elements with similar properties were grouped the same column. The columns in his table are like the piles of sorted objects in your junk drawer. Both contain groups of things with similar properties.

 Reading Check What physical property did Mendeleev use to place the elements in rows on the periodic table?

Patterns in Properties

The term *periodic* means "repeating pattern." For example, seasons and months are periodic because they follow a repeating pattern every year. The days of the week are periodic since they repeat every seven days.

What were some of the repeating patterns Mendeleev noticed in his table? Melting point is one property that shows a repeating pattern. Recall that melting point is the temperature at which a solid changes to a liquid. The blue line in **Figure 2** represents the melting points of the elements in row 2 of the periodic table. Notice that the melting point of carbon is higher than the melting point of lithium. However, the melting point of fluorine, at the far right of the row, is lower than that of carbon. How do these melting points show a pattern? Look at the red line in **Figure 2**. This line represents the melting points of the elements in row 3 of the periodic table. The melting points follow the same increasing and then decreasing pattern as the blue line, or row 2. Boiling point and reactivity also follow a periodic pattern.

A Periodic Property

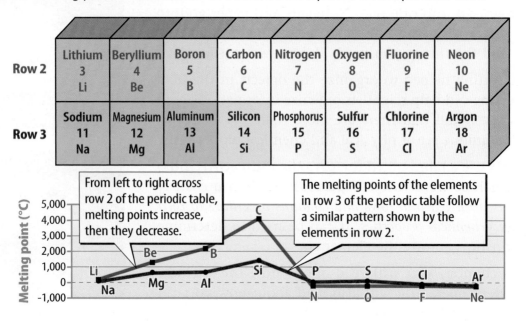

Figure 2 Melting points increase, then decrease, across a period on the periodic table.

346 • Chapter 10
EXPLAIN

Predicting Properties of Undiscovered Elements

When Mendeleev arranged all known elements by increasing atomic mass, there were large gaps between some elements. He predicted that scientists would discover elements that would fit into these spaces. Mendeleev also predicted that the properties of these elements would be similar to the known elements in the same columns. Both of his predictions turned out to be true.

Changes to Mendeleev's Table

Mendeleev's periodic table enabled scientists to relate the properties of the known elements to their position on the table. However, the table had a problem—some elements seemed out of place. Mendeleev believed that the atomic masses of certain elements must be invalid because the elements appeared in the wrong place on the periodic table. For example, Mendeleev placed tellurium before iodine despite the fact that tellurium has a greater atomic mass than iodine. He did so because iodine's properties more closely resemble those of fluorine and chlorine, just as copper's properties are closer to those of silver and gold, as shown in **Figure 3.**

FOLDABLES
Use four sheets of paper to make a top-tab book. Use it to organize your notes about the development of the periodic table.

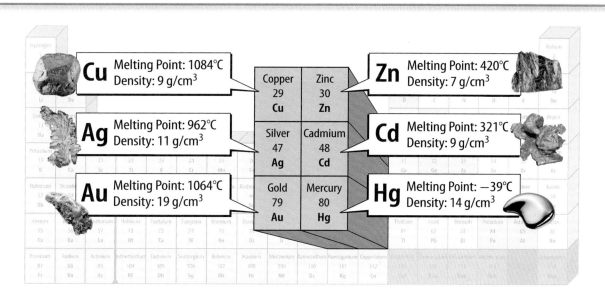

The Importance of Atomic Number

In the early 1900s, the scientist Henry Moseley solved the problem with Mendeleev's table. Moseley found that if elements were listed according to increasing atomic number instead of listing atomic mass, columns would contain elements with similar properties. Recall that the atomic number of an element is the number of protons in the nucleus of each of that element's atoms.

Figure 3 On today's periodic table, copper is in the same column as silver and gold. Zinc is in the same column as cadmium and mercury.

Concepts in Motion
Animation

 Key Concept Check What determines where an element is located on the periodic table you use today?

SCIENCE USE v. COMMON USE

period
Science Use the completion of a cycle; a row on the periodic table

Common Use a point used to mark the end of a sentence; a time frame

Today's Periodic Table

You can identify many of the properties of an element from its placement on the periodic table. The table, as shown in Figure 4, is organized into columns, rows, and blocks, which are based on certain patterns of properties. In the next two lessons, you will learn how an element's position on the periodic table can help you interpret the element's physical and chemical properties.

Figure 4 The periodic table is used to organize elements according to increasing atomic number and properties.

 Concepts in Motion   Animation

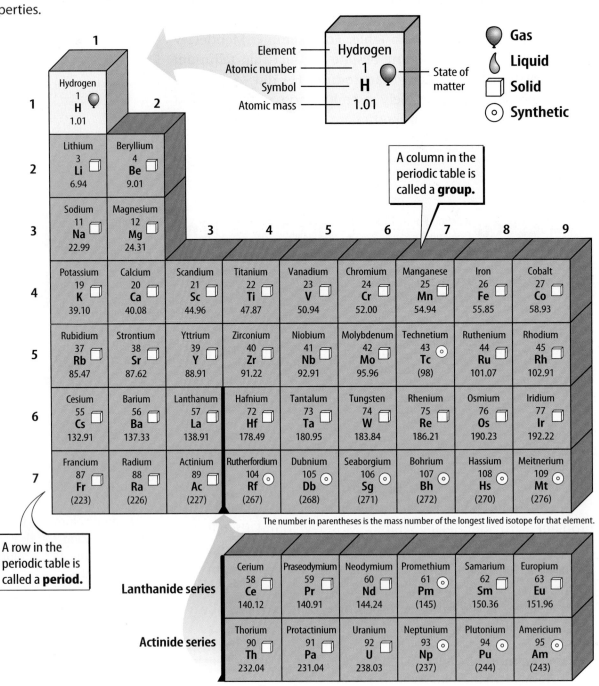

The number in parentheses is the mass number of the longest lived isotope for that element.

Chapter 10
EXPLAIN

What is on an element key?

The element key shows an element's chemical symbol, atomic number, and atomic mass. The key also contains a symbol that shows the state of matter at room temperature. Look at the element key for helium in **Figure 5**. Helium is a gas at room temperature. Some versions of the periodic table give additional information, such as density, conductivity, or melting point.

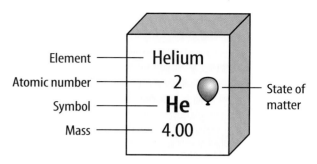

Figure 5 An element key shows important information about each element.

Visual Check What does this key tell you about helium?

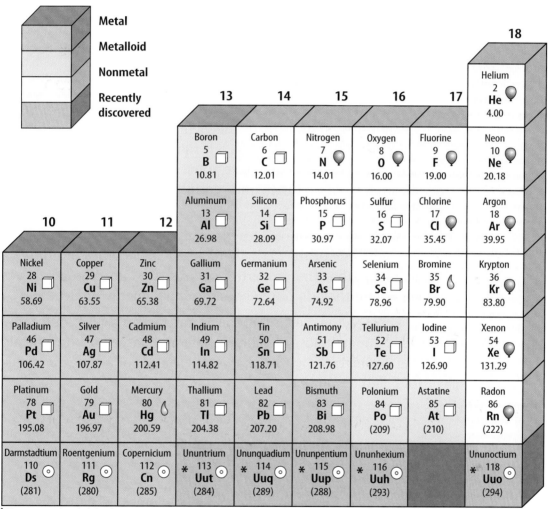

* The names and symbols for elements 113-116 and 118 are temporary. Final names will be selected when the elements' discoveries are verified.

Math Skills

Use Geometry
The distance around a circle is the circumference (C). The distance across the circle, through its center, is the diameter (d). The radius (r) is half of the diameter. The circumference divided by the diameter for any circle is equal to π (pi), or 3.14. The formula for determining the circumference is:

$C = \pi d$ or $C = 2\pi r$

For example, an iron (Fe) atom has a radius of **126 pm** (picometers; 1 picometer = one-trillionth of a meter) The circumference of an iron atom is:

$C = 2 \times 3.14 \times 126$ pm

$C = 791$ pm

Practice
The radius of a uranium (U) atom is 156 pm. What is its circumference?

- Math Practice
- Personal Tutor

Groups
A **group** *is a column on the periodic table.* Elements in the same group have similar chemical properties and react with other elements in similar ways. There are patterns in the physical properties of a group such as density, melting point, and boiling point. The groups are numbered 1–18, as shown in **Figure 4.**

 Key Concept Check What can you infer about the properties of two elements in the same group?

Periods
The rows on the periodic table are called **periods.** The atomic number of each element increases by one as you read from left to right across each period. The physical and chemical properties of the elements also change as you move left to right across a period.

Metals, Nonmetals, and Metalloids
Almost three-fourths of the elements on the periodic table are metals. Metals are on the left side and in the middle of the table. Individual metals have some properties that differ, but all metals are shiny and conduct thermal energy and electricity.

With the exception of hydrogen, nonmetals are located on the right side of the periodic table. The properties of nonmetals differ from the properties of metals. Many nonmetals are gases, and they do not conduct thermal energy or electricity.

Between the metals and the nonmetals on the periodic table are the metalloids. Metalloids have properties of both metals and nonmetals. **Figure 6** shows an example of a metal, a metalloid, and a nonmetal.

Figure 6 In period 3, magnesium is a metal, silicon is a metalloid, and sulfur is a nonmetal.

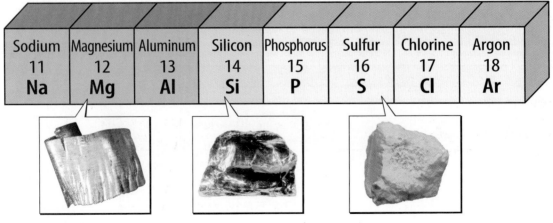

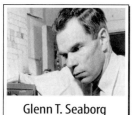

Glenn T. Seaborg — Seaborgium 106 Sg
Niels Bohr — Bohrium 107 Bh
Hassium 108 Hs
Lise Meitner — Meitnerium 109 Mt

Figure 7 Three of these synthetic elements are named to honor important scientists.

How Scientists Use the Periodic Table

Even today, new elements are created in laboratories, named, and added to the present-day periodic table. Four of these elements are shown in **Figure 7**. These elements are all synthetic, or made by people, and do not occur naturally on Earth. Sometimes scientists can create only a few atoms of a new element. Yet scientists can use the periodic table to predict the properties of new elements they create. Look back at the periodic table in **Figure 4**. What group would you predict to contain element 117? You would probably expect element 117 to be in group 17 and to have similar properties to other elements in the group. Scientists hope to one day synthesize element 117.

The periodic table contains more than 100 elements. Each element has unique properties that differ from the properties of other elements. But each element also shares similar properties with nearby elements. The periodic table shows how elements relate to each other and fit together into one organized chart. Scientists use the periodic table to understand and predict elements' properties. You can, too.

Reading Check How is the periodic table used to predict the properties of an element?

Inquiry MiniLab   20 minutes

How does atom size change across a period?
One pattern seen on the periodic table is in the radius of different atoms. The figure below shows how atomic radius is measured.

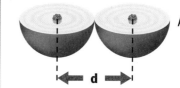

Atomic radius = $\frac{1}{2}d$

1. Read and complete a lab safety form.
2. Using **scissors** and **card stock paper,** cut seven 2-cm × 4-cm rectangles. Using a **marker,** label each rectangle with the atomic symbol of each of the first seven elements in period 2. Obtain the radius for each atom from your teacher.
3. Using a **ruler,** cut **plastic straws** to the same number of millimeters as each atomic radius given in picometers. For example, if the atomic radius is 145 pm, cut a straw 145 mm long.
4. **Tape** each of the labeled rectangles to the top of its appropriate straw.
5. Insert the straws into **modeling clay** according to increasing atomic number.

Analyze and Conclude
1. **Describe** the pattern you see in your model.
2. **Key Concept** Predict the pattern of atomic radii of the elements in period 4.

Lesson 1 EXPLAIN

Lesson 1 Review

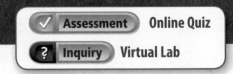

Visual Summary

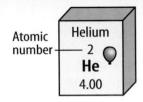

On the periodic table, elements are arranged according to increasing atomic number and similar properties.

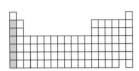

A column of the periodic table is called a group. Elements in the same group have similar properties.

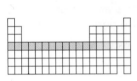

A row of the periodic table is called a period. Properties of elements repeat in the same pattern from left to right across each period.

FOLDABLES

Use your lesson Foldable to review the lesson. Save your Foldable for the project at the end of the chapter.

What do you think NOW?

You first read the statements below at the beginning of the chapter.

1. The elements on the periodic table are arranged in rows in the order they were discovered.
2. The properties of an element are related to the element's location on the periodic table.

Did you change your mind about whether you agree or disagree with the statements? Rewrite any false statements to make them true.

Use Vocabulary

1. **Identify** the scientific term used for rows on the periodic table.

2. **Name** the scientific term used for columns on the periodic table.

Understand Key Concepts

3. The _____ increases by one for each element as you move left to right across a period.

4. What does the decimal number in an element key represent?
   A. atomic mass   C. chemical symbol
   B. atomic number   D. state of matter

Interpret Graphics

5. **Classify** each marked element, 1 and 2, as a metal, a nonmetal, or a metalloid.

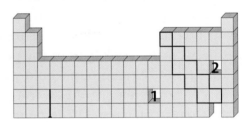

6. **Identify** Copy and fill in the graphic organizer below to identify the color-coded regions of the periodic table.

Critical Thinking

7. **Predict** Look at the perioidic table and predict three elements that have lower melting points than calcium (Ca).

Math Skills

— Math Practice —

8. Carbon (C) and silicon (Si) are in group 4 of the periodic table. The atomic radius of carbon is 77 pm and sulfur is 117 pm. What is the circumference of each atom?

Inquiry Skill Practice: Identify Patterns

25 minutes

How is the periodic table arranged?

Materials

20 cards

What would happen if schools did not assign students to grades or classes? How would you know where to go on the first day of school? What if your home did not have an address? How could you tell someone where you live? Life becomes easier with organization. The following activity will help you discover how elements are organized on the periodic table.

Learn It

Patterns help you make sense of the world around you. The days of the week follow a pattern, as do the months of the year. **Identifying a pattern** involves organizing things into similar groups and then sequencing the things in the same way in each group.

Try It

1. Obtain cards from your teacher. Turn the cards over so the sides with numbers are facing up.

2. Separate the cards into three or more piles. All of the cards in a pile should have a characteristic in common.

3. Organize each pile into a pattern. Use all of the cards.

4. Lay out the cards into rows and columns based on their characteristics and patterns.

Apply It

5. Describe in your Science Journal the patterns you used to organize your cards. Do other patterns exist in your arrangement?

6. Are there gaps in your arrangement? Can you describe what a card in one of those gaps would look like?

7. **Key Concept** What characteristics of elements might you use to organize them in a similar pattern?

Lesson 1
EXTEND
353

Lesson 2

Metals

Reading Guide

Key Concepts 🔑
ESSENTIAL QUESTIONS

- What elements are metals?
- What are the properties of metals?

Vocabulary
metal p. 355
luster p. 355
ductility p. 356
malleability p. 356
alkali metal p. 357
alkaline earth metal p. 357
transition element p. 358

 Multilingual eGlossary

Inquiry Where does it strike?

Lightning strikes the top of the Empire State Building approximately 100 times a year. Why does lightning hit the top of this building instead of the city streets or buildings below? Metal lightning rods allow electricity to flow through them more easily than other materials do. Lightning moves through these materials and the building is not harmed.

Launch Lab

20 minutes

What properties make metals useful?

The properties of metals determine their uses. Copper conducts thermal energy, which makes it useful for cookware. Aluminum has low density, so it is used in aircraft bodies. What other properties make metals useful?

1. Read and complete a lab safety form.
2. With your group, observe the **metal objects** in your **container.** For each object, discuss what properties allow the metal to be used in that way.
3. Observe the **photographs of gold and silver jewelry.** What properties make these two metals useful in jewelry?
4. Examine **other objects around the room** that you think are made of metal. Do they share the same properties as the objects in your container? Do they have other properties that make them useful?

Think About This

1. What properties do all the metals share? What properties are different?

2. **Key Concept** In your Science Journal, list at least four properties of metals that determine their uses.

What is a metal?

What do stainless steel knives and forks, copper wire, aluminum foil, and gold jewelry have in common? They are all made from metals.

As you read in Lesson 1, most of the elements on the periodic table are metals. In fact, of all the known elements, more than three-quarters are metals. With the exception of hydrogen, all of the elements in groups 1–12 on the periodic table are metals. In addition, some of the elements in groups 13–15 are metals. To be a metal, an element must have certain properties.

Key Concept Check How does the position of an element on the periodic table allow you to determine if the element is a metal?

Physical Properties of Metals

Recall that physical properties are characteristics used to describe or identify something without changing its makeup. All metals share certain physical properties.

A **metal** *is an element that is generally shiny. It is easily pulled into wires or hammered into thin sheets. A metal is a good conductor of electricity and thermal energy.* Gold exhibits the common properties of metals.

Luster and Conductivity People use gold for jewelry because of its beautiful color and metallic luster. **Luster** *describes the ability of a metal to reflect light.* Gold is also a good conductor of thermal energy and electricity. However, gold is too expensive to use in normal electrical wires or metal cookware. Copper is often used instead.

Lesson 2
EXPLORE

Properties of Metals

Figure 8 Gold has many uses based on its properties.

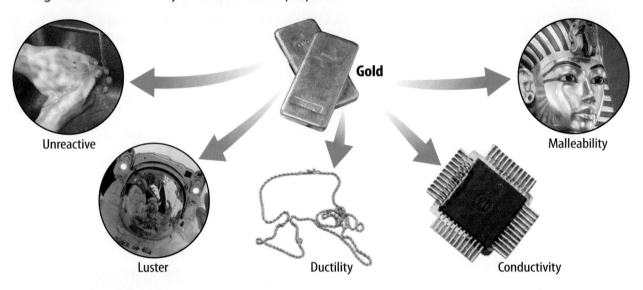

Unreactive · Luster · Gold · Ductility · Conductivity · Malleability

Visual Check Analyze why the properties shown in each photo are an advantage to using gold.

WORD ORIGIN

ductility
from Latin *ductilis*, means "may be led or drawn"

Ductility and Malleability Gold is the most ductile metal. **Ductility** (duk TIH luh tee) *is the ability to be pulled into thin wires.* A piece of gold with the mass of a paper clip can be pulled into a wire that is more than 3 km long.

Malleability (ma lee uh BIH luh tee) *is the ability of a substance to be hammered or rolled into sheets.* Gold is so malleable that it can be hammered into thin sheets. A pile of a million thin sheets would be only as high as a coffee mug.

REVIEW VOCABULARY

density
the mass per unit volume of a substance

Other Physical Properties of Metals In general the **density**, strength, boiling point, and melting point of a metal are greater than those of other elements. Except for mercury, all metals are solid at room temperature. Many uses of a metal are determined by the metal's physical properties, as shown in **Figure 8**.

Key Concept Check What are some physical properties of metals?

Chemical Properties of Metals

Recall that a chemical property is the ability or inability of a substance to change into one or more new substances. The chemical properties of metals can differ greatly. However, metals in the same group usually have similar chemical properties. For example, gold and other elements in group 11 do not easily react with other substances.

Make a two-tab book. Label it as shown. Use it to record information about the properties of metals.

356 Chapter 10
EXPLAIN

Group 1: Alkali Metals

The elements in group 1 are called **alkali** *(AL kuh li)* **metals.** The alkali metals include lithium, sodium, potassium, rubidium, cesium, and francium.

Because they are in the same group, alkali metals have similar chemical properties. Alkali metals react quickly with other elements, such as oxygen. Therefore, in nature, they occur only in compounds. Pure alkali metals must be stored so that they do not come in contact with oxygen and water vapor in the air. **Figure 9** shows potassium and sodium reacting with water.

Alkali metals also have similar physical properties. Pure alkali metals have a silvery appearance. As shown in **Figure 9,** they are soft enough to cut with a knife. The alkali metals also have the lowest densities of all metals. A block of pure sodium metal could float on water because of its very low density.

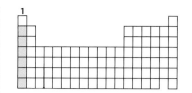

Figure 9 Alkali metals react violently with water. They are also soft enough to be cut with a knife.

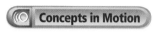
Animation

Potassium

Sodium

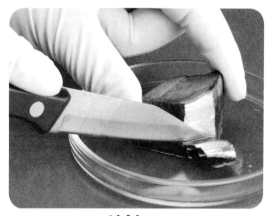
Lithium

Group 2: Alkaline Earth Metals

The elements in group 2 on the periodic table are called **alkaline** *(AL kuh lun)* **earth metals.** These metals are beryllium, magnesium, calcium, strontium, barium, and radium.

Alkaline earth metals also react quickly with other elements. However, they do not react as quickly as the alkali metals do. Like the alkali metals, pure alkaline earth metals do not occur naturally. Instead, they combine with other elements and form compounds. The physical properties of the alkaline earth metals are also similar to those of the alkali metals. Alkaline earth metals are soft and silvery. They also have low density, but they have greater density than alkali metals.

Reading Check Which element reacts faster with oxygen—barium or potassium?

Lesson 2
EXPLAIN

Transition Elements

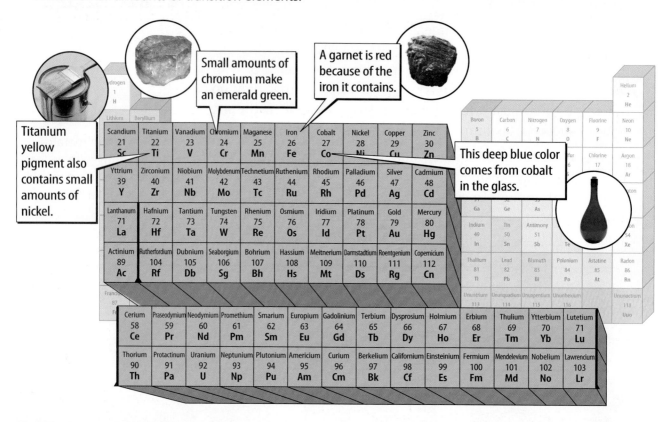

Figure 10 Transition elements are in blocks at the center of the periodic table. Many colorful materials contain small amounts of transition elements.

Titanium yellow pigment also contains small amounts of nickel.

Small amounts of chromium make an emerald green.

A garnet is red because of the iron it contains.

This deep blue color comes from cobalt in the glass.

Groups 3–12: Transition Elements

The elements in groups 3–12 are called **transition elements**. The transition elements are in two blocks on the periodic table. The main block is in the center of the periodic table. The other block includes the two rows at the bottom of the periodic table, as shown in **Figure 10**.

Properties of Transition Elements

All transition elements are metals. They have higher melting points, greater strength, and higher densities than the alkali metals and the alkaline earth metals. Transition elements also react less quickly with oxygen. Some transition elements can exist in nature as free elements. An element is a free element when it occurs in pure form, not in a compound.

Uses of Transition Elements

Transition elements in the main block of the periodic table have many important uses. Because of their high densities, strength, and resistance to corrosion, transition elements such as iron make good building materials. Copper, silver, nickel, and gold are used to make coins. These metals are also used for jewelry, electrical wires, and many industrial applications.

Main-block transition elements can react with other elements and form many compounds. Many of these compounds are colorful. Artists use transition-element compounds in paints and pigments. The color of many gems, such as garnets and emeralds, comes from the presence of small amounts of transition elements, as illustrated in **Figure 10**.

Lanthanide and Actinide Series

Two rows of transition elements are at the bottom of the periodic table, as shown in **Figure 10**. These elements were removed from the main part of the table so that periods 6 and 7 were not longer than the other periods. If these elements were included in the main part of the table, the first row, called the lanthanide series, would stretch between lanthanum and halfnium. The second row, called the actinide series, would stretch between actinium and rutherfordium.

Some lanthanide and actinide series elements have valuable properties. For example, lanthanide series elements are used to make strong magnets. Plutonium, one of the actinide series elements, is used as a fuel in some nuclear reactors.

Patterns in Properties of Metals

Recall that the properties of elements follow repeating patterns across the periods of the periodic table. In general, elements increase in metallic properties such as luster, malleability, and electrical conductivity from right to left across a period, as shown in **Figure 11**. The elements on the far right of a period have no metallic properties at all. Potassium (K), the element on the far left in period 4, has the highest luster, is the most malleable, and conducts electricity better than all the elements in this period.

There are also patterns within groups. Metallic properties tend to increase as you move down a group, also shown in **Figure 11**. You could predict that the malleability of gold is greater than the malleability of either silver or copper because it is below these two elements in group 11.

Reading Check Where would you expect to find elements on the periodic table with few or no metallic properties?

Inquiry MiniLab — 20 minutes

How well do materials conduct thermal energy?

How well a material conducts thermal energy can often determine its use.

1. Read and complete a lab safety form.
2. Have your teacher add about 200 mL of very **hot water** to a **250-mL beaker**.
3. Place **rods of metal, plastic, glass, and wood** in the water for 30 seconds.
4. Set four large **ice cubes** on a sheet of **paper towel**. Use **tongs** to quickly remove each rod from the hot water. Place the heated end of the rod on an ice cube.
5. After 30 seconds, remove the rods and examine the ice cubes.

Analyze and Conclude

1. **Conclude** What can you conclude about how well metals conduct thermal energy?
2. **Key Concept** Cookware is often made of metal. What property of metals makes them useful for this purpose?

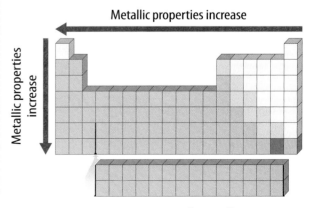

Figure 11 Metallic properties of elements increase as you move to the left and down on the periodic table.

Lesson 2 Review

Assessment Online Quiz

Visual Summary

Properties of metals include conductivity, luster, malleability, and ductility.

Alkali metals and alkaline earth metals react easily with other elements. These metals make up groups 1 and 2 on the periodic table.

Transition elements make up groups 3–12 and the lanthanide and actinide series on the periodic table.

FOLDABLES

Use your lesson Foldable to review the lesson. Save your Foldable for the project at the end of the chapter.

What do you think NOW?

You first read the statements below at the beginning of the chapter.

3. Fewer than half of the elements are metals.

4. Metals are usually good conductors of electricity.

Did you change your mind about whether you agree or disagree with the statements? Rewrite any false statements to make them true.

Use Vocabulary

1 **Use the term** *luster* in a sentence.

2 **Identify** the property that makes copper metal ideal for wiring.

3 Elements that have the lowest densities of all the metals are called _____.

Understand Key Concepts

4 **List** the physical properties that most metals have in common.

5 Which is a chemical property of transition elements?
  A. brightly colored
  B. great ductility
  C. denser than alkali metals
  D. reacts little with oxygen

6 **Organize** the following metals from least metallic to most metallic: barium, zinc, iron, and strontium.

Interpret Graphics

7 **Examine** this section of the periodic table. What metal will have properties most similar to those of chromium (Cr)? Why?

Vanadium 23 V	Chromium 24 Cr	Maganese 25 Mn
Niobium 41 Nb	Molybdenum 42 Mo	Technetium 43 Tc

Critical Thinking

8 **Investigate** your classroom and locate five examples of materials made from metal.

9 **Evaluate** the physical properties of potassium, magnesium, and copper. Select the best choice to use for a building project. Explain why this metal is the best building material to use.

Fireworks

SCIENCE & SOCIETY

Metals add a variety of colors to fireworks.

About 1,000 years ago, the Chinese discovered the chemical formula for gunpowder. Using this formula, they invented the first fireworks. One of the primary ingredients in gunpowder is saltpeter, or potassium nitrate. Find potassium on the periodic table. Notice that potassium is a metal. How does the chemical behavior of a metal contribute to a colorful fireworks show?

Purple: mix of strontium and copper compounds

Blue: copper compounds

Yellow: sodium compounds

Gold: iron burned with carbon

White-hot: barium-oxygen compounds or aluminum or magnesium burn

Orange: calcium compounds

Green: barium compounds

Red: strontium and lithium compounds

Metal compounds contribute to the variety of colors you see at a fireworks show. Recall that metals have special chemical and physical properties. Compounds that contain metals also have special properties. For example, each metal turns a characteristic color when burned. Lithium, an alkali metal, forms compounds that burn red. Copper compounds burn blue. Aluminum and magnesium burn white.

It's Your Turn

FORM AN OPINION Fireworks contain metal compounds. Are they bad for the environment or your health? Research the effects of metals on human health and on the environment. Decide if fireworks are safe to use for holiday celebrations.

Lesson 3

Nonmetals and Metalloids

Reading Guide

Key Concepts
ESSENTIAL QUESTIONS

- Where are nonmetals and metalloids on the periodic table?
- What are the properties of nonmetals and metalloids?

Vocabulary

nonmetal p. 363
halogen p. 365
noble gas p. 366
metalloid p. 367
semiconductor p. 367

 Multilingual eGlossary

Inquiry Why don't they melt?

What do you expect to happen to something when a flame is placed against it? As you can see, the nonmetal material this flower sits on protects the flower from the flame. Some materials conduct thermal energy. Other materials, such as this one, do not.

362 Chapter 10
ENGAGE

Launch Lab

20 minutes

What are some properties of nonmetals?

You now know what the properties of metals are. What properties do nonmetals have?

1. Read and complete a lab safety form.
2. Examine pieces of **copper, carbon, aluminum,** and **sulfur.** Describe the appearance of these elements in your Science Journal.
3. Use a **conductivity tester** to check how well these elements conduct electricity. Record your observations.
4. Wrap each element sample in a **paper towel.** Carefully hit the sample with a **hammer.** Unwrap the towel and observe the sample. Record your observations.

Think About This

1. Locate these elements on the periodic table. From their locations, which elements are metals? Which elements are nonmetals?
2. **Key Concept** Using your results, compare the properties of metals and nonmetals.
3. **Key Concept** What property of a nonmetal makes it useful to insulate electrical wires?

The Elements of Life

Would it surprise you to learn that more than 96 percent of the mass of your body comes from just four elements? As shown in **Figure 12,** all four of these elements—oxygen, carbon, hydrogen, and nitrogen—are nonmetals. **Nonmetals** *are elements that have no metallic properties.*

Of the remaining elements in your body, the two most common elements also are nonmetals—phosphorus and sulfur. These six elements form the compounds in proteins, fats, nucleic acids, and other large molecules in your body and in all other living things.

Reading Check What are the six most common elements in the human body?

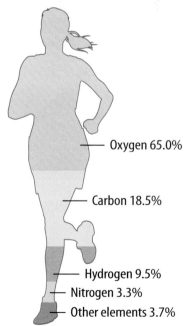

Figure 12 Like other living things, this woman's mass comes mostly from nonmetals.

- Oxygen 65.0%
- Carbon 18.5%
- Hydrogen 9.5%
- Nitrogen 3.3%
- Other elements 3.7%

Lesson 3
EXPLORE

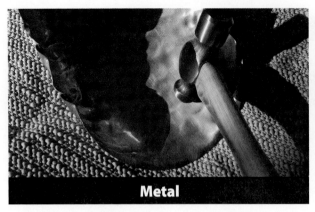

▲ **Figure 13** Solid metals, such as copper, are malleable. Solid nonmetals, such as sulfur, are brittle.

How are nonmetals different from metals?

Recall that metals have luster. They are ductile, malleable, and good conductors of electricity and thermal energy. All metals except mercury are solids at room temperature.

The properties of nonmetals are different from those of metals. Many nonmetals are gases at room temperature. Those that are solid at room temperature have a dull surface, which means they have no luster. Because nonmetals are poor conductors of electricity and thermal energy, they are good insulators. For example, nose cones on space shuttles are insulated from the intense thermal energy of reentry by a material made from carbon, a nonmetal. **Figure 13** and **Figure 14** show several properties of nonmetals.

Figure 14 Nonmetals have properties that are different from those of metals. Phosphorus and carbon are dull, brittle solids that do not conduct thermal energy or electricity. ▼

Key Concept Check What properties do nonmetals have?

Properties of Nonmetals

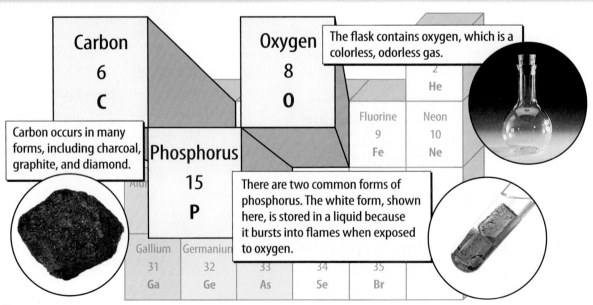

Visual Check Compare the properties of oxygen to those of carbon and phosphorus.

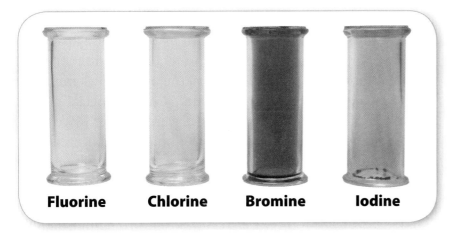

Figure 15 These glass containers each hold a halogen gas. Although they are different colors in their gaseous state, they react similarly with other elements.

Visual Check Compare the colors of these halogens.

Nonmetals in Groups 14–16

Look back at the periodic table in **Figure 4.** Notice that groups 14–16 contain metals, nonmetals, and metalloids. The chemical properties of the elements in each group are similar. However, the physical properties of the elements can be quite different.

Carbon is the only nonmetal in group 14. It is a solid that has different forms. Carbon is in most of the compounds that make up living things. Nitrogen, a gas, and phosphorus, a solid, are the only nonmetals in group 15. These two elements form many different compounds with other elements, such as oxygen. Group 16 contains three nonmetals. Oxygen is a gas that is essential for many organisms. Sulfur and selenium are solids that have the physical properties of other solid nonmetals.

Group 17: The Halogens

An element in group 17 of the periodic table is called a **halogen** (HA luh jun). **Figure 15** shows the halogens fluorine, chlorine, bromine, and iodine. The term *halogen* refers to an element that can react with a metal and form a salt. For example, chlorine gas reacts with solid sodium and forms sodium chloride, or table salt. Calcium chloride is another salt often used on icy roads.

Halogens react readily with other elements and form compounds. They react so readily that halogens only can occur naturally in compounds. They do not exist as free elements. They even form compounds with other nonmetals, such as carbon. In general, the halogens are less reactive as you move down the group.

Reading Check Will bromine react with sodium? Explain your answer.

FOLDABLES
Fold a sheet of paper to make a table with three columns and three rows. Label it as shown. Use it to organize information about nonmetals and metalloids.

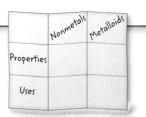

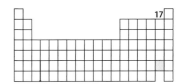

WORD ORIGIN
halogen
from Greek *hals*, means "salt"; and *–gen,* means "to produce"

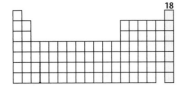

Group 18: The Noble Gases

The elements in group 18 are known as the **noble gases.** The elements helium, neon, argon, krypton, xenon, and radon are the noble gases. Unlike the halogens, the only way elements in this group react with other elements is under special conditions in a laboratory. These elements were not yet discovered when Mendeleev constructed his periodic table because they do not form compounds naturally. Once they were discovered, they fit into a group at the far right side of the table.

Hydrogen

Figure 16 shows the element key for hydrogen. Of all the elements, hydrogen has the smallest atomic mass. It is also the most common element in the universe.

Is hydrogen a metal or a nonmetal? Hydrogen is most often classified as a nonmetal because it has many properties like those of nonmetals. For example, like some nonmetals, hydrogen is a gas at room temperature. However, hydrogen also has some properties similar to those of the group 1 alkali metals. In its liquid form, hydrogen conducts electricity just like a metal does. In some chemical reactions, hydrogen reacts as if it were an alkali metal. However, under conditions on Earth, hydrogen usually behaves like a nonmetal.

 Reading Check Why is hydrogen usually classified as a nonmetal?

ACADEMIC VOCABULARY
construct
(verb) to make by combining and arranging parts

Figure 16 More than 90 percent of all the atoms in the universe are hydrogen atoms. Hydrogen is the main fuel for the nuclear reactions that occur in stars.

Metalloids

Between the metals and the nonmetals on the periodic table are elements known as metalloids. *A* **metalloid** *(MEH tul oyd) is an element that has physical and chemical properties of both metals and nonmetals.* The elements boron, silicon, germanium, arsenic, antimony, tellurium, polonium, and astatine are metalloids. Silicon is the most abundant metalloid in the universe. Most sand is made of a compound containing silicon. Silicon is also used in many different products, some of which are shown in **Figure 17.**

Key Concept Check Where are metalloids on the periodic table?

Semiconductors

Recall that metals are good conductors of thermal energy and electricity. Nonmetals are poor conductors of thermal energy and electricity but are good insulators. A property of metalloids is the ability to act as a semiconductor. *A* **semiconductor** *conducts electricity at high temperatures, but not at low temperatures.* At high temperatures, metalloids act like metals and conduct electricity. But at lower temperatures, metalloids act like nonmetals and stop electricity from flowing. This property is useful in electronic devices such as computers, televisions, and solar cells.

WORD ORIGIN

semiconductor
from Latin *semi-*, means "half"; and *conducere*, means "to bring together"

Figure 17 The properties of silicon make it useful for many different products.

Uses of Silicon

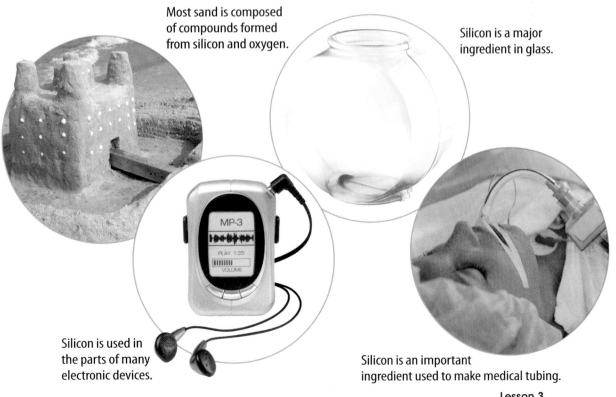

Most sand is composed of compounds formed from silicon and oxygen.

Silicon is a major ingredient in glass.

Silicon is used in the parts of many electronic devices.

Silicon is an important ingredient used to make medical tubing.

Figure 18 This microchip conducts electricity at high temperatures using a semiconductor.

Properties and Uses of Metalloids

Pure silicon is used in making semiconductor devices for computers and other electronic products. Germanium is also used as a semiconductor. However, metalloids have other uses, as shown in **Figure 18.** Pure silicon and Germanium are used in semiconductors. Boron is used in water softeners and laundry products. Boron also glows bright green in fireworks. Silicon is one of the most abundant elements on Earth. Sand, clay, and many rocks and minerals are made of silicon compounds.

Metals, Nonmetals, and Metalloids

You have read that all metallic elements have common characteristics, such as malleability, conductivity, and ductility. However, each metal has unique properties that make it different from other metals. The same is true for nonmetals and metalloids. How can knowing the properties of an element help you evaluate its uses?

Look again at the periodic table. An element's position on the periodic table tells you a lot about the element. By knowing that sulfur is a nonmetal, for example, you know that it breaks easily and does not conduct electricity. You would not choose sulfur to make a wire. You would not try to use oxygen as a semiconductor or sodium as a building material. You know that transition elements are strong, malleable, and do not react easily with oxygen or water. Because of these characteristics, these metals make good building materials. Understanding the properties of elements can help you decide which element to use in a given situation.

Reading Check Why would you not use an element on the right side of the periodic table as a building material?

Inquiry MiniLab   15 minutes

Which insulates better?

In this lab, you will compare how well a metal bowl and a nonmetal ball containing a mixture of nonmetals conduct thermal energy.

1. Read and complete a lab safety form.
2. Pour **very warm water** into a **pitcher.**
3. Pour half of the warm water into a **metal bowl.** In your Science Journal, describe how the outside of the bowl feels.
4. Inflate a **beach ball** until it is one-third full. Mold the partially filled beach ball into the shape of a bowl. Pour the remaining warm water into your beach ball bowl. Feel the outside of the bowl. Describe how it feels.

Analyze and Conclude

1. **Explain** the difference in the outside temperatures of the two bowls.
2. **Predict** the results of putting ice in each of the bowls.
3. **Key Concept** Make a statement about how well a nonmetal conducts thermal energy.

Lesson 3 Review

Assessment Online Quiz

Visual Summary

A nonmetal is an element that has no metallic properties. Solid nonmetals are dull, brittle, and do not conduct thermal energy or electricity.

Halogens and noble gases are nonmetals. These elements are found in group 17 and group 18 of the periodic table.

Metalloids have some metallic properties and some nonmetallic properties. The most important use of metalloids is as semiconductors.

FOLDABLES

Use your lesson Foldable to review the lesson. Save your Foldable for the project at the end of the chapter.

What do you think NOW?

You first read the statements below at the beginning of the chapter.

5. Most of the elements in living things are nonmetals.

6. Even though they look very different, oxygen and sulfur share some similar properties.

Did you change your mind about whether you agree or disagree with the statements? Rewrite any false statements to make them true.

Use Vocabulary

1 Distinguish between a nonmetal and a metalloid.

2 An element in group 17 of the periodic table is called a(n) _____.

3 An element in group 18 of the periodic table is called a(n) _____.

Understand Key Concepts

4 The ability of a halogen to react with a metal to form a salt is an example of a _____ property.
  A. chemical   C. periodic
  B. noble gas   D. physical

5 Classify each of the following elements as a metal, a nonmetal, or a metalloid: boron, carbon, aluminum, and silicon.

6 Infer which group you would expect to contain element 117. Use the periodic table to help you answer this question.

Interpret Graphics

7 Sequence nonmetals, metals, and metalloids in order from left to right across the periodic table by copying and completing the graphic organizer below.

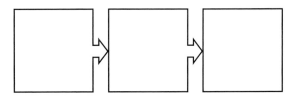

Critical Thinking

8 Hypothesize how your classroom would be different if there were no metalloids.

9 Analyze why hydrogen is sometimes classified as a metal.

10 Determine whether there would be more nonmetals in group 14 or in group 16. Explain your answer.

Lesson 3
EVALUATE
369

Alien Insect Periodic Table

Materials

cards

The periodic table classifies elements according to their properties. In this lab, you will model the procedure used to develop the periodic table. Your model will include developing patterns using pictures of alien insects. You will then use your patterns to predict what missing alien insects look like.

Question

How can I arrange objects into patterns by using their properties?

Procedure

1. Obtain a set of alien insect pictures. Spread them out so you can see all of them. Observe the pictures with a partner. Look for properties that you might use to organize the pictures.

2. Make a list of properties you might use to group the alien insects. These properties are those that a number of insects have in common.

3. Make a list of properties you might use to sequence the insects. These properties change from one insect to the next in some pattern.

4. With your partner, decide what pattern you will use to arrange the alien insects in an organized rectangular block. All the insects in a vertical column, or group, must be the same in some way. They must also share some feature that changes regularly as you move down the group. All the aliens in a horizontal row, or period, must be the same in some way and must also share some feature that changes regularly as you move across the period.

5. Arrange your insects as you planned. Two insects are missing from your set. Leave empty spaces in your rectangular block for these pictures. When you have finished arranging your insects, have the teacher check your pattern.

6. Write a description of the properties you predict each missing alien insect will have. Then draw a picture of each missing insect.

Analyze and Conclude

7. **Explain** Could you have predicted the properties of the missing insects without placing the others in a pattern? Why or why not?

8. **The Big Idea** How is your arrangement similar to the one developed by Mendeleev for elements? How is it different?

9. **Infer** What properties can you use to predict the identity of one missing insect? What do you not know about that insect?

Lab Tips

☑ A property is any observable characteristic that you can use to distinguish between objects.

☑ A pattern is a consistent plan or model used as a guide for understanding or predicting something.

Communicate Your Results

Create a slide show presentation that demonstrates, step by step, how you grouped and sequenced your insects and predicted the properties of the missing insects. Show your presentation to students in another class.

How could you change the insects so that they better represent the properties of elements, such as atomic mass?

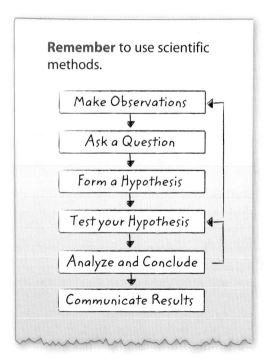

Remember to use scientific methods.

- Make Observations
- Ask a Question
- Form a Hypothesis
- Test your Hypothesis
- Analyze and Conclude
- Communicate Results

Chapter 10 Study Guide

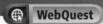

Elements are organized on the periodic table according to increasing atomic number and similar properties.

Key Concepts Summary

Lesson 1: Using the Periodic Table
- Elements are organized on the **periodic table** by increasing atomic number and similar properties.
- Elements in the same **group**, or column, of the periodic table have similar properties.
- Elements' properties change across a **period**, which is a row of the periodic table.
- Each element key on the periodic table provides the name, symbol, atomic number, and atomic mass for an element.

Vocabulary
periodic table p. 345
group p. 350
period p. 350

Lesson 2: Metals
- **Metals** are located on the left and middle side of the periodic table.
- Metals are elements that have **ductility, malleability, luster,** and conductivity.
- The **alkali metals** are in group 1 of the periodic table, and the **alkaline earth metals** are in group 2.
- **Transition elements** are metals in groups 3–12 of the periodic table, as well as the lanthanide and actinide series.

metal p. 355
luster p. 355
ductility p. 356
malleability p. 356
alkali metal p. 357
alkaline earth metal p. 357
transition element p. 358

Lesson 3: Nonmetals and Metalloids
- **Nonmetals** are on the right side of the periodic table, and **metalloids** are located between metals and nonmetals.
- Nonmetals are elements that have no metallic properties. Solid nonmetals are dull in appearance, brittle, and do not conduct electricity. Metalloids are elements that have properties of both metals and nonmetals.
- Some metalloids are **semiconductors.**
- Elements in group 17 are called **halogens,** and elements in group 18 are **noble gases.**

nonmetal p. 363
halogen p. 365
noble gas p. 366
metalloid p. 367
semiconductor p. 367

Study Guide

- Personal Tutor
- Vocabulary eGames
- Vocabulary eFlashcards

Chapter Project

Assemble your lesson Foldables as shown to make a Chapter Project. Use the project to review what you have learned in this chapter.

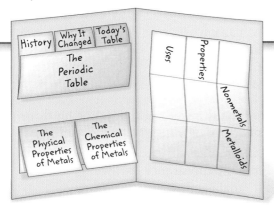

Use Vocabulary

1. The element magnesium (Mg) is in _____ 3 of the periodic table.

2. An element that is shiny, is easily pulled into wires or hammered into thin sheets, and is a good conductor of electricity and heat is a(n) _____.

3. Copper is used to make wire because it has the property of _____.

4. An element that is sometimes a good conductor of electricity and sometimes a good insulator is a(n) _____.

5. An element that is a poor conductor of heat and electricity but is a good insulator is a(n) _____.

Link Vocabulary and Key Concepts

Interactive Concept Map

Copy this concept map, and then use vocabulary terms from the previous page to complete the concept map.

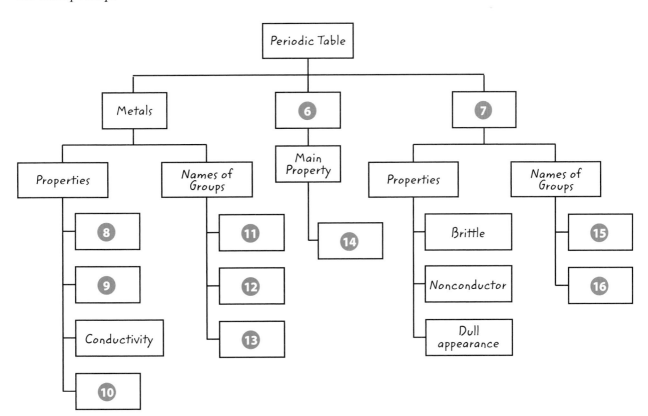

Chapter 10 Study Guide • 373

Chapter 10 Review

Understand Key Concepts

1. What determines the order of elements on today's periodic table?
   A. increasing atomic mass
   B. decreasing atomic mass
   C. increasing atomic number
   D. decreasing atomic number

2. The element key for nitrogen is shown below.

From this key, determine the atomic mass of nitrogen.
   A. 7
   B. 7.01
   C. 14.01
   D. 21.01

3. Look at the periodic table in Lesson 1. Which of the following lists of elements forms a group on the periodic table?
   A. Li, Be, B, C, N, O, F, and Ne
   B. He, Ne, Ar, Kr, Xe, and Rn
   C. B, Si, As, Te, and At
   D. Sc, Ti, V, Cr, Mn, Fe, Co, Cu, Ni, and Zn

4. Which is NOT a property of metals?
   A. brittleness
   B. conductivity
   C. ductility
   D. luster

5. What are two properties that make a metal a good choice to use as wire in electronics?
   A. conductivity, malleability
   B. ductility, conductivity
   C. luster, malleability
   D. malleability, high density

6. Where are most metals on the periodic table?
   A. on the left side only
   B. on the right side only
   C. in the middle only
   D. on the left side and in the middle

7. Look at the periodic table in Lesson 1 and determine which element is a metalloid.
   A. carbon
   B. silicon
   C. oxygen
   D. aluminum

8. Iodine is a solid nonmetal. What is one property of iodine?
   A. conductivity
   B. dull appearance
   C. malleability
   D. ductility

9. The following table lists some information about certain elements in group 17.

Element Symbol	Atomic Number	Melting Point (°C)	Boiling Point (°C)
F	9	−233	−187
Cl	17	−102	−35
Br	35	−7.3	59
I	53	114	183

Which statement describes what happens to these elements as atomic number increases?
   A. Both melting point and boiling point decrease.
   B. Melting point increases and boiling point decreases.
   C. Melting point decreases and boiling point increases.
   D. Both melting point and boiling point increase.

374 • Chapter 10 Review

Chapter Review

Assessment — Online Test Practice

Critical Thinking

10 Recommend an element to use to fill bottles that contain ancient paper. The element should be a gas at room temperature, should be denser than helium, and should not easily react with other elements.

11 Apply Why is mercury the only metal to have been used in thermometers?

12 Evaluate the following types of metals as a choice to make a Sun reflector: alkali metals, alkaline earth metals, or transition metals. The metal cannot react with water or oxygen and must be shiny and strong.

13 The figure below shows a pattern of densities.

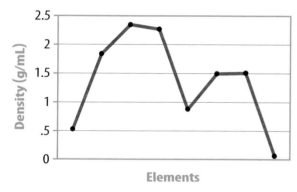

Infer whether you are looking at a graph of elements within a group or across a period. Explain your answer.

14 Contrast aluminum and nitrogen. Show why aluminum is a metal and nitrogen is not.

15 Classify A student sorted six elements. He placed iron, silver, and sodium in group A. He placed neon, oxygen, and nitrogen in group B. Name one other element that fits in group A and another element that belongs in group B. Explain your answer.

Writing in Science

16 Write a plan that shows how a metal, a nonmetal, and a metalloid could be used when constructing a building.

REVIEW THE BIG IDEA

17 Explain how atomic number and properties are used to determine where element 115 is placed on the periodic table.

18 The photo below shows how the properties of materials determine their uses. How can the periodic table be used to help you find elements with properties similar to that of helium?

Math Skills

Review — Math Practice

Use Geometry

19 The table below shows the atomic radii of three elements in group 1 on the periodic table.

Element	Atomic radius
Li	152 pm
Na	186 pm
K	227 pm

a. What is the circumference of each atom?

b. Rubidium (Rb) is the next element in Group 1. What would you predict about the radius and circumference of a rubidium atom?

Chapter 10 Review • 375

Standardized Test Practice

Record your answers on the answer sheet provided by your teacher or on a sheet of paper.

Multiple Choice

1 Where are most nonmetals located on the periodic table?

 A in the bottom row
 B on the left side and in the middle
 C on the right side
 D in the top row

Use the figure below to answer question 2.

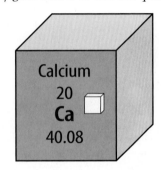

2 What is the atomic mass of calcium?

 A 20
 B 40.08
 C 40.08 ÷ 20
 D 40.08 + 20

3 Which element is most likely to react with potassium?

 A bromine
 B calcium
 C nickel
 D sodium

4 Which group of elements can act as semiconductors?

 A halogens
 B metalloids
 C metals
 D noble gases

Use the table below about group 13 elements to answer question 5.

Element Symbol	Atomic Number	Density (g/cm^3)	Atomic Mass
B	5	2.34	10.81
Al	13	2.70	26.98
Ga	31	5.90	69.72
In	49	7.30	114.82

5 How do density and atomic mass change as atomic number increases?

 A Density and atomic mass decrease.
 B Density and atomic mass increase.
 C Density decreases and atomic mass increases.
 D Density increases and atomic mass decreases.

6 Which elements have high densities, strength, and resistance to corrosion?

 A alkali metals
 B alkaline earth metals
 C metalloids
 D transition elements

7 Which is a property of a metal?

 A It is brittle.
 B It is a good insulator.
 C It has a dull appearance.
 D It is malleable.

Standardized Test Practice

Use the figure below to answer questions 8 and 9.

8 The figure shows a group in the periodic table. What is the name of this group of elements?

  A halogens
  B metalloids
  C metals
  D noble gases

9 Which is a property of these elements?

  A They are conductors.
  B They are semiconductors.
  C They are nonreactive with other elements.
  D They react easily with other elements.

10 What is one similarity among elements in a group?

  A atomic mass
  B atomic weight
  C chemical properties
  D practical uses

Constructed Response

Use the figure below to answer questions 11 and 12.

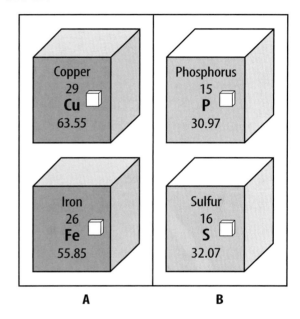

11 Groups A and B each contain two elements. Identify each group as metals, nonmetals, or metalloids. Would silicon belong to one of these groups? Why or why not?

12 Which group in the figure above yields the strongest building elements? Why?

13 How does the periodic table of elements help scientists today?

14 What connection does the human body have with the elements on the periodic table?

NEED EXTRA HELP?														
If You Missed Question...	1	2	3	4	5	6	7	8	9	10	11	12	13	14
Go to Lesson...	1	1	3	3	1	2	2	3	3	1	2,3	2	1	3

Chapter 11

Elements and Chemical Bonds

THE BIG IDEA How do elements join together to form chemical compounds?

Inquiry How do they combine?

How many different words could you type using just the letters on a keyboard? The English alphabet has only 26 letters, but a dictionary lists hundreds of thousands of words using these letters! Similarly only about 115 different elements make all kinds of matter.

- How do so few elements form so many different kinds of matter?
- Why do you think different types of matter have different properties?
- How are atoms held together to produce different types of matter?

Get Ready to Read

What do you think?

Before you read, decide if you agree or disagree with each of these statements. As you read this chapter, see if you change your mind about any of the statements.

1. Elements rarely exist in pure form. Instead, combinations of elements make up most of the matter around you.

2. Chemical bonds that form between atoms involve electrons.

3. The atoms in a water molecule are more chemically stable than they would be as individual atoms.

4. Many substances dissolve easily in water because opposite ends of a water molecule have opposite charges.

5. Losing electrons can make some atoms more chemically stable.

6. Metals are good electrical conductors because they tend to hold onto their valence electrons very tightly.

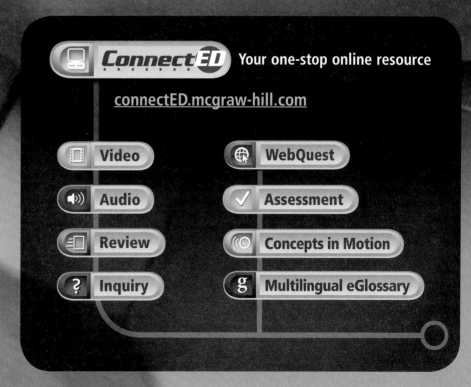

ConnectED Your one-stop online resource

connectED.mcgraw-hill.com

- Video
- Audio
- Review
- Inquiry
- WebQuest
- Assessment
- Concepts in Motion
- Multilingual eGlossary

Lesson 1

Reading Guide

Key Concepts
ESSENTIAL QUESTIONS

- How is an electron's energy related to its distance from the nucleus?
- Why do atoms gain, lose, or share electrons?

Vocabulary
chemical bond p. 382
valence electron p. 384
electron dot diagram p. 385

 Multilingual eGlossary

 Video BrainPOP®

Electrons and Energy Levels

Inquiry Are pairs more stable?

Rowing can be hard work, especially if you are part of a racing team. The job is made easier because the rowers each pull on the water with a pair of oars. How do pairs make the boat more stable?

Launch Lab

20 minutes

How is the periodic table organized?

How do you begin to put together a puzzle of a thousand pieces? You first sort similar pieces into groups. All edge pieces might go into one pile. All blue pieces might go into another pile. Similarly, scientists placed the elements into groups based on their properties. They created the periodic table, which organizes information about all the elements.

1. Obtain six **index cards** from your teacher. Using one card for each element name, write the names *beryllium, sodium, iron, zinc, aluminum,* and *oxygen* at the top of a card.

2. Open your textbook to the periodic table printed on the inside back cover. Locate the element key for each element written on your cards.

3. For each element, find the following information and write it on the index card: symbol, atomic number, atomic mass, state of matter, and element type.

Think About This

1. What do the elements in the blue blocks have in common? In the green blocks? In the yellow blocks?

2. **Key Concept** Each element in a column on the periodic table has similar chemical properties and forms bonds in similar ways. Based on this, for each element you listed on a card, name another element on the periodic table that has similar chemical properties.

The Periodic Table

Imagine trying to find a book in a library if all the books were unorganized. Books are organized in a library to help you easily find the information you need. The periodic table is like a library of information about all chemical elements.

A copy of the periodic table is on the inside back cover of this book. The table has more than 100 blocks—one for each known element. Each block on the periodic table includes basic properties of each element such as the element's state of matter at room temperature and its atomic number. The atomic number is the number of protons in each atom of the element. Each block also lists an element's atomic mass, or the average mass of all the different isotopes of that element.

Periods and Groups

You can learn about some properties of an element from its position on the periodic table. Elements are organized in periods (rows) and groups (columns). The periodic table lists elements in order of atomic number. The atomic number increases from left to right as you move across a period. Elements in each group have similar chemical properties and react with other elements in similar ways. In this lesson, you will read more about how an element's position on the periodic table can be used to predict its properties.

Reading Check How is the periodic table organized?

Lesson 1

EXPLORE 381

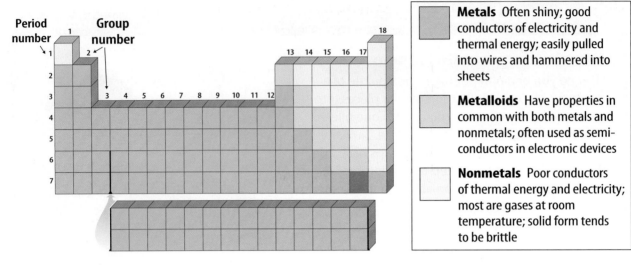

▲ **Figure 1** Elements on the periodic table are classified as metals, nonmetals, or metalloids.

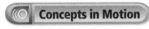
Animation

REVIEW VOCABULARY

compound
matter that is made up of two or more different kinds of atoms joined together by chemical bonds

Figure 2 Protons and neutrons are in an atom's nucleus. Electrons move around the nucleus. ▼

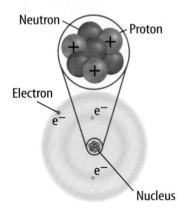

Metals, Nonmetals, and Metalloids

The three main regions of elements on the periodic table are shown in **Figure 1.** Except for hydrogen, elements on the left side of the table are metals. Nonmetals are on the right side of the table. Metalloids form the narrow stair-step region between metals and nonmetals.

Reading Check Where are metals, nonmetals, and metalloids on the periodic table?

Atoms Bond

In nature, pure elements are rare. Instead, atoms of different elements chemically combine and form **compounds.** Compounds make up most of the matter around you, including living and nonliving things. There are only about 115 elements, but these elements combine and form millions of compounds. Chemical bonds hold them together. *A* **chemical bond** *is a force that holds two or more atoms together.*

Electron Number and Arrangement

Recall that atoms contain protons, neutrons, and electrons, as shown in **Figure 2.** Each proton has a positive charge; each neutron has no charge; and each electron has a negative charge. The atomic number of an element is the number of protons in each atom of that element. In a neutral (uncharged) atom, the number of protons equals the number of electrons.

The exact position of electrons in an atom cannot be determined. This is because electrons are in constant motion around the nucleus. However, each electron is usually in a certain area of space around the nucleus. Some are in areas close to the nucleus, and some are in areas farther away.

Electrons and Energy Different electrons in an atom have different amounts of energy. An electron moves around the nucleus at a distance that corresponds to its amount of energy. Areas of space in which electrons move around the nucleus are called energy levels. Electrons closest to the nucleus have the least amount of energy. They are in the lowest energy level. Electrons farthest from the nucleus have the greatest amount of energy. They are in the highest energy level. The energy levels of an atom are shown in **Figure 3.** Notice that only two electrons can be in the lowest energy level. The second energy level can hold up to eight.

 Key Concept Check How is an electron's energy related to its position in an atom?

Electrons and Bonding Imagine two magnets. The closer they are to each other, the stronger the attraction of their opposite ends. Negatively charged electrons have a similar attraction to the positively charged nucleus of an atom. The electrons in energy levels closest to the nucleus of the same atom have a strong attraction to that nucleus. However, electrons farther from that nucleus are weakly attracted to it. These outermost electrons can easily be attracted to the nucleus of other atoms. This attraction between the positive nucleus of one atom and the negative electrons of another is what causes a chemical bond.

FOLDABLES
Make two quarter-sheet note cards from a sheet of paper. Use them to organize your notes on valence electrons and electron dot diagrams.

Figure 3 Electrons are in certain energy levels within an atom.

 Review Personal Tutor

Electron Energy Levels

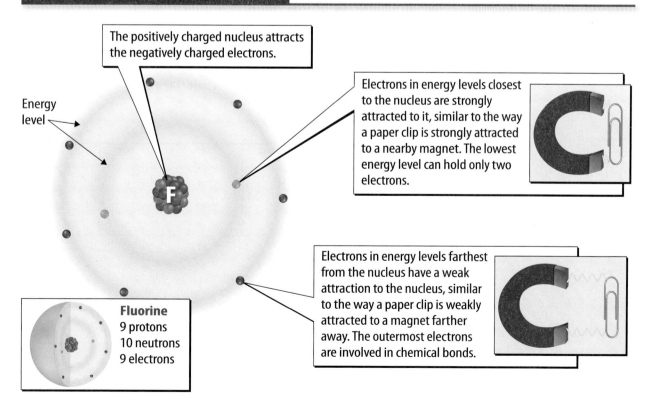

Valence Electrons

You have read that electrons farthest from their nucleus are easily attracted to the nuclei of nearby atoms. These outermost electrons are the only electrons involved in chemical bonding. Even atoms that have only a few electrons, such as hydrogen or lithium, can form chemical bonds. This is because these electrons are still the outermost electrons and are exposed to the nuclei of other atoms. A **valence electron** *is an outermost electron of an atom that participates in chemical bonding.* Valence electrons have the most energy of all electrons in an atom.

The number of valence electrons in each atom of an element can help determine the type and the number of bonds it can form. How do you know how many valence electrons an atom has? The periodic table can tell you. Except for helium, elements in certain groups have the same number of valence electrons. **Figure 4** illustrates how to use the periodic table to determine the number of valence electrons in the atoms of groups 1, 2, and 13–18. Determining the number of valence electrons for elements in groups 3–12 is more complicated. You will learn about these groups in later chemistry courses.

> **Word Origin**
> **valence**
> from Latin *valentia*, means "strength, capacity"

Figure 4 You can use the group numbers at the top of the columns to determine the number of valence electrons in atoms of groups 1, 2, and 13–18.

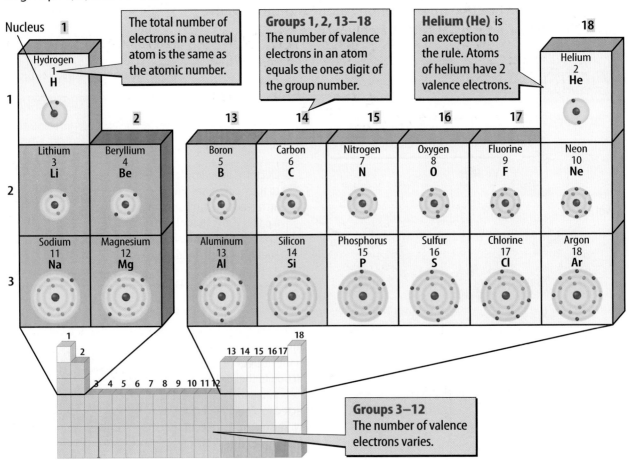

Visual Check How many valence electrons does an atom of phosphorous (P) have?

Writing and Using Electron Dot Diagrams

Figure 5 Electron dot diagrams show the number of valence electrons in an atom.

Steps for writing a dot diagram	Beryllium	Carbon	Nitrogen	Argon
1 Identify the element's group number on the periodic table.	2	14	15	18
2 Identify the number of valence electrons. • This equals the ones digit of the group number.	2	4	5	8
3 Draw the electron dot diagram. • Place one dot at a time on each side of the symbol (top, right, bottom, left). Repeat until all dots are used.	Be·	·C̈·	·N̈·	:Är:
4 Determine if the atom is chemically stable. • An atom is chemically stable if all dots on the electron dot diagram are paired.	Chemically Unstable	Chemically Unstable	Chemically Unstable	Chemically Stable
5 Determine how many bonds this atom can form. • Count the dots that are unpaired.	2	4	3	0

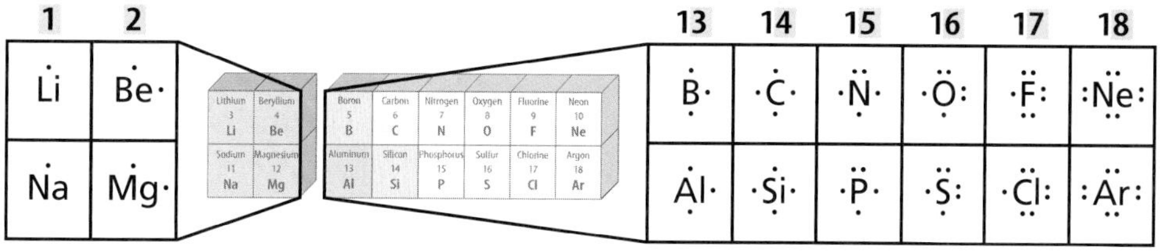

Electron Dot Diagrams

In 1916 an American Chemist named Gilbert Lewis developed a method to show an element's valence electrons. He developed the **electron dot diagram**, *a model that represents valence electrons in an atom as dots around the element's chemical symbol.*

Electron dot diagrams can help you predict how an atom will bond with other atoms. Dots, representing valence electrons, are placed one-by-one on each side of an element's chemical symbol until all the dots are used. Some dots will be paired up, others will not. The number of unpaired dots is often the number of bonds an atom can form. The steps for writing dot diagrams are shown in **Figure 5**.

Reading Check Why are electron dot diagrams useful?

Recall that each element in a group has the same number of valence electrons. As a result, every element in a group has the same number of dots in its electron dot diagram.

Notice in **Figure 5** that an argon atom, Ar, has eight valence electrons, or four pairs of dots, in the diagram. There are no unpaired dots. Atoms with eight valence electrons do not easily react with other atoms. They are chemically stable. Atoms that have between one and seven valence electrons are reactive, or chemically unstable. These atoms easily bond with other atoms and form chemically stable compounds.

Atoms of hydrogen and helium have only one energy level. These atoms are chemically stable with two valence electrons.

Lesson 1
EXPLAIN

Inquiry MiniLab

20 minutes

How does an electron's energy relate to its position in an atom?

Electrons in energy levels closest to the nucleus are strongly attracted to it. You can use paper clips and a magnet to model a similar attraction.

1. Read and complete a lab safety form.
2. Pick up a **paper clip** with a **magnet**. Use the first paper clip to pick up another one.
3. Continue picking up paper clips in this way until you have a chain of paper clips and no more will attach.
4. Gently pull off the paper clips one by one.

Analyze and Conclude

1. **Observe** Which paper clip was the easiest to remove? Which was the most difficult?
2. **Use Models** In what way do the magnet and the paper clips act as a model for an atom?
3. **Key Concept** How does an electron's position in an atom affect its ability to take part in chemical bonding?

Noble Gases

The elements in Group 18 are called noble gases. With the exception of helium, noble gases have eight valence electrons and are chemically stable. Chemically stable atoms do not easily react, or form bonds, with other atoms. The electron structures of two noble gases—neon and helium—are shown in **Figure 6.** Notice that all dots are paired in the dot diagrams of these atoms.

Stable and Unstable Atoms

Atoms with unpaired dots in their electron dot diagrams are reactive, or chemically unstable. For example, nitrogen, shown in **Figure 6,** has three unpaired dots in its electron dot diagram, and it is reactive. Nitrogen, like many other atoms, becomes more stable by forming chemical bonds with other atoms.

When an atom forms a bond, it gains, loses, or shares valence electrons with other atoms. By forming bonds, atoms become more chemically stable. Recall that atoms are most stable with eight valence electrons. Therefore, atoms with less than eight valence electrons form chemical bonds and become stable. In Lessons 2 and 3, you will read which atoms gain, lose, or share electrons when forming stable compounds.

Key Concept Check Why do atoms gain, lose, or share electrons?

Figure 6 Atoms gain, lose, or share valence electrons and become chemically stable.

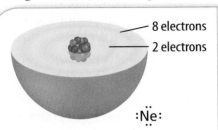

Neon has 10 electrons: 2 inner electrons and 8 valence electrons. A neon atom is chemically stable because it has 8 valence electrons. All dots in the dot diagram are paired.

Helium has 2 electrons. Because an atom's lowest energy level can hold only 2 electrons, the 2 dots in the dot diagram are paired. Helium is chemically stable.

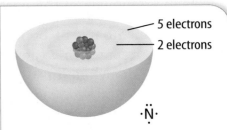

Nitrogen has 7 electrons: 2 inner electrons and 5 valence electrons. Its dot diagram has 1 pair of dots and 3 unpaired dots. Nitrogen atoms become more stable by forming chemical bonds.

Lesson 1 Review

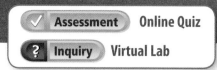

Visual Summary

Electrons are less strongly attracted to a nucleus the farther they are from it, similar to the way a magnet attracts a paper clip.

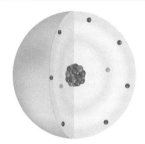

Electrons in atoms are in energy levels around the nucleus. Valence electrons are the outermost electrons.

All noble gases, except He, have four pairs of dots in their electron dot diagrams. Noble gases are chemically stable.

FOLDABLES

Use your lesson Foldable to review the lesson. Save your Foldable for the project at the end of the chapter.

What do you think NOW?

You first read the statements below at the beginning of the chapter.

1. Elements rarely exist in pure form. Instead, combinations of elements make up most of the matter around you.
2. Chemical bonds that form between atoms involve electrons.

Did you change your mind about whether you agree or disagree with the statements? Rewrite any false statements to make them true.

Use Vocabulary

1. **Use the term** *chemical bond* in a complete sentence.

2. **Define** *electron dot diagram* in your own words.

3. The electrons of an atom that participate in chemical bonding are called _____.

Understand Key Concepts

4. **Identify** the number of valence electrons in each atom: calcium, carbon, and sulfur.

5. Which part of the atom is shared, gained, or lost when forming a chemical bond?
   A. electron
   B. neutron
   C. nucleus
   D. proton

6. **Draw** electron dot diagrams for oxygen, potassium, iodine, nitrogen, and beryllium.

Interpret Graphics

7. **Determine** the number of valence electrons in each diagram shown below.

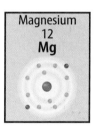
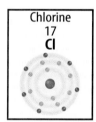

8. **Organize Information** Copy and fill in the graphic organizer below to describe one or more details for each concept: electron energy, valence electrons, stable atoms.

Concept	Description

Critical Thinking

9. **Compare** krypton and bromine in terms of chemical stability.

10. **Decide** An atom of nitrogen has five valence electrons. How could a nitrogen atom become more chemically stable?

GREEN SCIENCE

New Green Airships

The Difference of One Valence Electron

Faster than ocean liners and safer than airplanes, airships used to be the best way to travel. The largest, the *Hindenburg*, was nearly the size of the *Titanic*. To this day, no larger aircraft has ever flown. So, what happened to the giant airship? The answer lies in a valence electron.

The builders of the *Hindenburg* filled it with a lighter-than-air gas, hydrogen, so that it would float. Their plan was to use helium, a noble gas. However, helium was scarce. They knew hydrogen was explosive, but it was easier to get. For nine years, hydrogen airships floated safely back and forth across the Atlantic. But in 1937, disaster struck. Just before it landed, the *Hindenburg* exploded in flames. The age of the airship was over.

Since the *Hindenburg*, airplanes have become the main type of air transportation. A big airplane uses hundreds of gallons of fuel to take off and fly. As a result, it releases large amounts of pollutants into the atmosphere. Some people are looking for other types of air transportation that will be less harmful to the environment. Airships may be the answer. An airship floats and needs very little fuel to take off and stay airborne. Airships also produce far less pollution than other aircraft.

Today, however, airships use helium not hydrogen. With two valence electrons instead of one, as hydrogen has, helium is unreactive. Thanks to helium's chemical stability, someday you might be a passenger on a new, luxurious, but not explosive, version of the *Hindenburg*.

▲ A new generation of big airships might soon be hauling freight and carrying passengers.

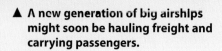

RESEARCH Precious documents deteriorate with age as their surfaces react with air. Parchment turns brown and crumbles. Find out how our founding documents have been saved from this fate by noble gases.

Lesson 2

Reading Guide

Key Concepts 🔑
ESSENTIAL QUESTIONS

- How do elements differ from the compounds they form?
- What are some common properties of a covalent compound?
- Why is water a polar compound?

Vocabulary
covalent bond p. 391
molecule p. 392
polar molecule p. 393
chemical formula p. 394

g Multilingual eGlossary

Compounds, Chemical Formulas, and Covalent Bonds

Inquiry How do they combine?

A jigsaw puzzle has pieces that connect in a certain way. The pieces fit together by sharing tabs with other pieces. All of the pieces combine and form a complete puzzle. Like pieces of a puzzle, atoms can join together and form a compound by sharing electrons.

Inquiry Launch Lab

20 minutes

How is a compound different from its elements?

The sugar you use to sweeten foods at home is probably sucrose. Sucrose contains the elements carbon, hydrogen, and oxygen. How does table sugar differ from the elements that it contains?

1. Read and complete a lab safety form.
2. Air is a mixture of several gases, including oxygen and hydrogen. Charcoal is a form of carbon. Write some properties of oxygen, hydrogen, and carbon in your Science Journal.
3. Obtain from your teacher a piece of **charcoal** and a **beaker** with **table sugar** in it.
4. Observe the charcoal. In your Science Journal, describe the way it looks and feels.
5. Observe the table sugar in the beaker. What does it look and feel like? Record your observations.

Think About This

1. Compare and contrast the properties of charcoal, hydrogen, and oxygen.
2. **Key Concept** How do you think the physical properties of carbon, hydrogen, and oxygen change when they combined to form sugar?

From Elements to Compounds

Have you ever baked cupcakes? First, combine flour, baking soda, and a pinch of salt. Then, add sugar, eggs, vanilla, milk, and butter. Each ingredient has unique physical and chemical properties. When you mix the ingredients together and bake them, a new product results—cupcakes. The cupcakes have properties that are different from the ingredients.

In some ways, compounds are like cupcakes. Recall that a compound is a substance made up of two or more different elements. Just as cupcakes are different from their ingredients, compounds are different from their elements. An element is made of one type of atom, but compounds are chemical combinations of different types of atoms. Compounds and the elements that make them up often have different properties.

Chemical **bonds** join atoms together. Recall that a chemical bond is a force that holds atoms together in a compound. In this lesson, you will learn that one way that atoms can form bonds is by sharing valence electrons. You will also learn how to write and read a chemical formula.

Key Concept Check How is a compound different from the elements that compose it?

SCIENCE USE V. COMMON USE

bond

Science Use a force that holds atoms together in a compound

Common Use a close personal relationship between two people

390 Chapter 11
EXPLORE

Covalent Bonds

Figure 7 A covalent bond forms when two nonmetal atoms share electrons.

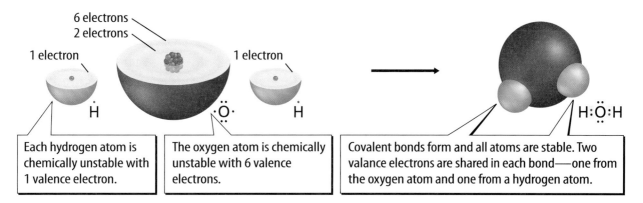

Each hydrogen atom is chemically unstable with 1 valence electron.

The oxygen atom is chemically unstable with 6 valence electrons.

Covalent bonds form and all atoms are stable. Two valance electrons are shared in each bond—one from the oxygen atom and one from a hydrogen atom.

Covalent Bonds—Electron Sharing

As you read in Lesson 1, one way that atoms can become more chemically stable is by sharing valence electrons. When unstable, nonmetal atoms bond together, they bond by sharing valence electrons. *A **covalent bond** is a chemical bond formed when two atoms share one or more pairs of valence electrons.* The atoms then form a stable covalent compound.

A Noble Gas Electron Arrangement

Look at the reaction between hydrogen and oxygen in **Figure 7.** Before the reaction, each hydrogen atom has one valence electron. The oxygen atom has six valence electrons. Recall that most atoms are chemically stable with eight valence electrons—the same electron arrangement as a noble gas. An atom with less than eight valence electrons becomes stable by forming chemical bonds until it has eight valence electrons. Therefore, an oxygen atom forms two bonds to become stable. A hydrogen atom is stable with two valence electrons. It forms one bond to become stable.

Shared Electrons

If the oxygen atom and each hydrogen atom share their unpaired valence electrons, they can form two covalent bonds and become a stable covalent compound. Each covalent bond contains two valence electrons—one from the hydrogen atom and one from the oxygen atom. Since these electrons are shared, they count as valence electrons for both atoms in the bond. Each hydrogen atom now has two valence electrons. The oxygen atom now has eight valence electrons, since it bonds to two hydrogen atoms. All three atoms have the electron arrangement of a noble gas and the compound is stable.

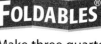

Make three quarter-sheet note cards from a sheet of paper to organize information about single, double, and triple covalent bonds.

Lesson 2
EXPLAIN

Double and Triple Covalent Bonds

As shown in **Figure 8**, a single covalent bond exists when two atoms share one pair of valence electrons. A double covalent bond exists when two atoms share two pairs of valence electrons. Double bonds are stronger than single bonds. A triple covalent bond exists when two atoms share three pairs of valence electrons. Triple bonds are stronger than double bonds. Multiple bonds are explained in **Figure 8**.

Covalent Compounds

When two or more atoms share valence electrons, they form a stable covalent compound. The covalent compounds carbon dioxide, water, and sugar are very different, but they also share similar properties. Covalent compounds usually have low melting points and low boiling points. They are usually gases or liquids at room temperature, but they can also be solids. Covalent compounds are poor conductors of thermal energy and electricity.

Molecules

The chemically stable unit of a covalent compound is a molecule. *A* **molecule** *is a group of atoms held together by covalent bonding that acts as an independent unit.* Table sugar ($C_{12}H_{22}O_{11}$) is a covalent compound. One grain of sugar is made up of trillions of sugar molecules. Imagine breaking a grain of sugar into the tiniest microscopic particle possible. You would have a molecule of sugar. One sugar molecule contains 12 carbon atoms, 22 hydrogen atoms, and 11 oxygen atoms all covalently bonded together. The only way to further break down the molecule would be to chemically separate the carbon, hydrogen, and oxygen atoms. These atoms alone have very different properties from the compound sugar.

 Key Concept Check What are some common properties of covalent compounds?

Multiple Bonds

Figure 8 The more valence electrons that two atoms share, the stronger the covalent bond is between the atoms.

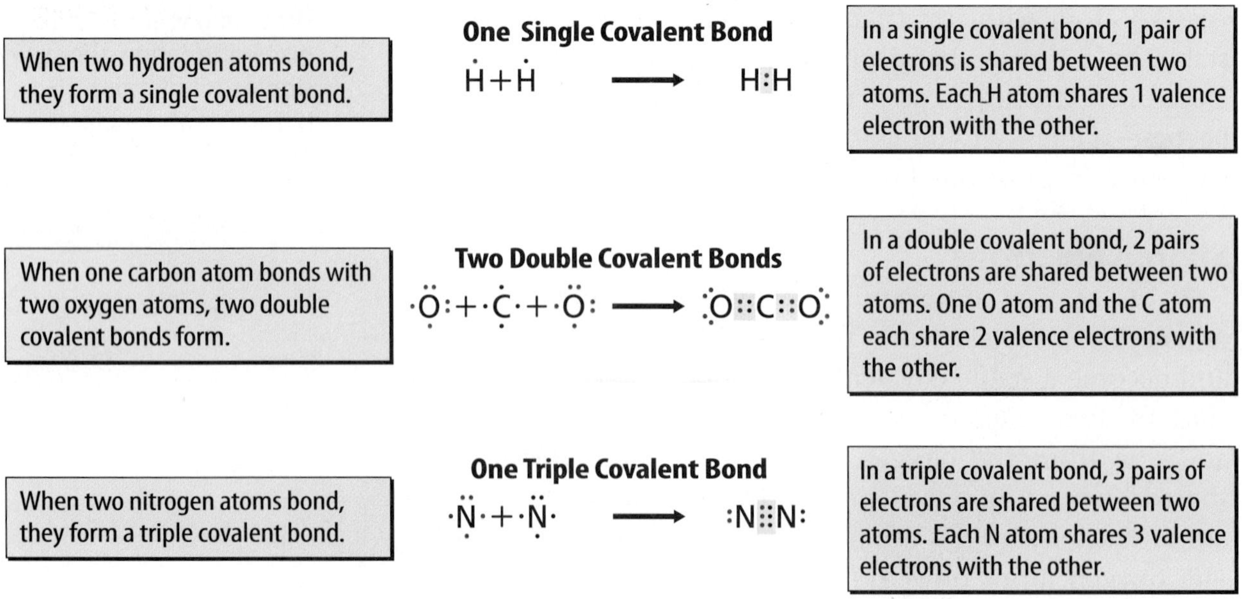

	One Single Covalent Bond	
When two hydrogen atoms bond, they form a single covalent bond.	Ḣ + Ḣ ⟶ H:H	In a single covalent bond, 1 pair of electrons is shared between two atoms. Each H atom shares 1 valence electron with the other.
	Two Double Covalent Bonds	
When one carbon atom bonds with two oxygen atoms, two double covalent bonds form.	·Ö: + ·Ċ· + ·Ö: ⟶ :Ö::C::Ö:	In a double covalent bond, 2 pairs of electrons are shared between two atoms. One O atom and the C atom each share 2 valence electrons with the other.
	One Triple Covalent Bond	
When two nitrogen atoms bond, they form a triple covalent bond.	·Ṅ· + ·Ṅ· ⟶ :N⋮⋮N:	In a triple covalent bond, 3 pairs of electrons are shared between two atoms. Each N atom shares 3 valence electrons with the other.

Visual Check Is the bond stronger between atoms in hydrogen gas (H_2) or nitrogen gas (N_2)? Why?

Water and Other Polar Molecules

In a covalent bond, one atom can attract the shared electrons more strongly than the other atom can. Think about the valence electrons shared between oxygen and hydrogen atoms in a water molecule. The oxygen atom attracts the shared electrons more strongly than each hydrogen atom does. As a result, the shared electrons are pulled closer to the oxygen atom, as shown in **Figure 9.** Since electrons have a negative charge, the oxygen atom has a partial negative charge. The hydrogen atoms have a partial positive charge. *A molecule that has a partial positive end and a partial negative end because of unequal sharing of electrons is a* **polar molecule.**

The charges on a polar molecule affect its properties. Sugar, for example, dissolves easily in water because both sugar and water are polar. The negative end of a water molecule pulls on the positive end of a sugar molecule. Also, the positive end of a water molecule pulls on the negative end of a sugar molecule. This causes the sugar molecules to separate from one another and mix with the water molecules.

 Key Concept Check Why is water a polar compound?

Nonpolar Molecules

A hydrogen molecule, H_2, is a nonpolar molecule. Because the two hydrogen atoms are identical, their attraction for the shared electrons is equal. The carbon dioxide molecule, CO_2, in **Figure 9** is also nonpolar. A nonpolar compound will not easily dissolve in a polar compound, but it will dissolve in other nonpolar compounds. Oil is an example of a nonpolar compound. It will not dissolve in water. Have you ever heard someone say "like dissolves like"? This means that polar compounds can dissolve in other polar compounds. Similarly, nonpolar compounds can dissolve in other nonpolar compounds.

WORD ORIGIN
polar
from Latin *polus*, means "pole"

Figure 9 Atoms of a polar molecule share their valence electrons unequally. Atoms of a nonpolar molecule share their valence electrons equally.

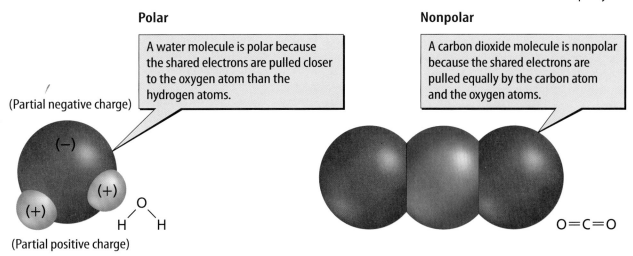

Inquiry MiniLab
20 minutes

How do compounds form?
Use building blocks to model ways in which elements combine to form compounds.

1. Examine various types of **interlocking plastic blocks.** Notice that the blocks have different numbers of holes and pegs. Attaching one peg to one hole represents a shared pair of electrons.

2. Draw the electron dot diagrams for carbon, nitrogen, oxygen, and hydrogen in your Science Journal. Based on the diagrams, decide which block should represent an atom of each element.

3. Use the blocks to make models of H_2, CO_2, NH_3, H_2O, and CH_4. All pegs on the largest block must fit into a hole, and no blocks can stick out over the edge of a block, either above or below it.

Analyze and Conclude
1. **Explain** how you decided which type of block should be assigned to each type of atom.
2. **Key Concept** Name at least one way that your models show the difference between a compound and the elements that combine and form the compound.

Chemical Formulas and Molecular Models

How do you know which elements make up a compound? *A **chemical formula** is a group of chemical symbols and numbers that represent the elements and the number of atoms of each element that make up a compound.* Just as a recipe lists ingredients, a chemical formula lists the elements in a compound. For example, the chemical formula for carbon dioxide shown in **Figure 10** is CO_2. The formula uses chemical symbols that show which elements are in the compound. Notice that CO_2 is made up of carbon (C) and oxygen (O). A subscript, or small number after a chemical symbol, shows the number of atoms of each element in the compound. Carbon dioxide (CO_2) contains two atoms of oxygen bonded to one atom of carbon.

A chemical formula describes the types of atoms in a compound or a molecule, but it does not explain the shape or appearance of the molecule. There are many ways to model a molecule. Each one can show the molecule in a different way. Common types of models for CO_2 are shown in **Figure 10**.

Reading Check What information is given in a chemical formula?

Figure 10 Chemical formulas and molecular models provide information about molecules.

Chemical Formula
A carbon dioxide molecule is made up of carbon (C) and oxygen (O) atoms.

CO_2

A symbol without a subscript indicates one atom. Each molecule of carbon dioxide has one carbon atom.

The subscript 2 indicates two atoms of oxygen. Each molecule of carbon dioxide has two oxygen atoms.

Dot Diagram
- Shows atoms and valence electrons

$\ddot{\text{O}}::\text{C}::\ddot{\text{O}}$

Structural Formula
- Shows atoms and lines; each line represents one shared pair of electrons

O=C=O

Ball-and-Stick Model
- Balls represent atoms and sticks represent bonds; used to show bond angles

Space-Filling Model
- Spheres represent atoms; used to show three-dimensional arrangement of atoms

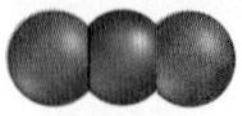

Lesson 2 Review

Assessment | Online Quiz

Visual Summary

CO₂ — A chemical formula is one way to show the elements that make up a compound.

O=C=O — A covalent bond forms when atoms share valence electrons. The smallest particle of a covalent compound is a molecule.

Water is a polar molecule because the oxygen and hydrogen atoms unequally share electrons.

FOLDABLES

Use your lesson Foldable to review the lesson. Save your Foldable for the project at the end of the chapter.

What do you think NOW?

You first read the statements below at the beginning of the chapter.

3. The atoms in a water molecule are more chemically stable than they would be as individual atoms.

4. Many substances dissolve easily in water because opposite ends of a water molecule have opposite charges.

Did you change your mind about whether you agree or disagree with the statements? Rewrite any false statements to make them true.

Use Vocabulary

1. **Define** *covalent bond* in your own words.

2. The group of symbols and numbers that shows the types and numbers of atoms that make up a compound is a _____.

3. **Use the term** *molecule* in a complete sentence.

Understand Key Concepts

4. **Contrast** Name at least one way water (H_2O) is different from the elements that make up water.

5. **Explain** why water is a polar molecule.

6. A sulfur dioxide molecule has one sulfur atom and two oxygen atoms. Which is its correct chemical formula?
   A. SO_2
   B. $(SO)_2$
   C. S_2O_2
   D. S_2O

Interpret Graphics

7. **Examine** the electron dot diagram for chlorine below.

In chlorine gas, two chlorine atoms join to form a Cl_2 molecule. How many pairs of valence electrons do the atoms share?

8. **Compare and Contrast** Copy and fill in the graphic organizer below to identify at least one way polar and nonpolar molecules are similar and one way they are different.

Polar and Nonpolar Molecules	
Similarities	
Differences	

Critical Thinking

9. **Develop** an analogy to explain the unequal sharing of valence electrons in a water molecule.

Lesson 2
EVALUATE
395

Inquiry Skill Practice — Model

25 minutes

How can you model compounds?

Chemists use models to explain how electrons are arranged in an atom. Electron dot diagrams are models used to show how many valence electrons an atom has. Electron dot diagrams are useful because they can help predict the number and type of bond an atom will form.

Materials

colored pencils

Learn It

In science, **models** are used to help you visualize objects that are too small, too large, or too complex to understand. A model is a representation of an object, idea, or event.

Try It

1. Use the periodic table to write the electron dot diagrams for hydrogen, oxygen, carbon, and silicon.

2. Using your electron dot diagrams from step 1, write electron dot diagrams for the following compounds: H_2O, CO, CO_2, SiO_2, C_2H_2, and CH_4. Use colored pencils to differentiate the electrons for each atom. Remember that all the above atoms, except hydrogen and helium, are chemically stable when they have eight valence electrons. Hydrogen and helium are chemically stable with two valence electrons.

Apply It

3. Based on your model, describe silicon's electron dot diagram and arrangement of valence electrons before and after it forms the compound SiO_2.

4. 🔑 **Key Concept** Which of the covalent compounds you modeled contain double bonds? Which contain triple bonds?

396 • Chapter 11
EXTEND

Lesson 3

Reading Guide

Key Concepts
ESSENTIAL QUESTIONS

- What is an ionic compound?
- How do metallic bonds differ from covalent and ionic bonds?

Vocabulary
ion p. 398
ionic bond p. 400
metallic bond p. 401

g Multilingual eGlossary

Ionic and Metallic Bonds

Inquiry) What is this?

This scene might look like snow along a shoreline, but it is actually thick deposits of salt on a lake. Over time, tiny amounts of salt dissolved in river water that flowed into this lake and built up as water evaporated. Salt is a compound that forms when elements form bonds by gaining or losing valence electrons, not sharing them.

Inquiry Launch Lab

15 minutes

How can atoms form compounds by gaining and losing electrons?

Metals often lose electrons when forming stable compounds. Nonmetals often gain electrons.

1. Read and complete a lab safety form.
2. Make two model atoms of sodium, and one model atom each of calcium, chlorine, and sulfur. To do this, write each element's chemical symbol with a **marker** on a **paper plate.** Surround the symbol with small balls of **clay** to represent valence electrons. Use one color of clay for the metals (groups 1 and 2 elements) and another color of clay for nonmetals (groups 16 and 17 elements).
3. To model sodium sulfide (Na_2S), place the two sodium atoms next to the sulfur atom. To form a stable compound, move each sodium atom's valence electron to the sulfur atom.
4. Form as many other compound models as you can by removing valence electrons from the groups 1 and 2 plates and placing them on the groups 16 and 17 plates.

Think About This

1. What other compounds were you able to form?
2. **Key Concept** How do you think your models are different from covalent compounds?

FOLDABLES

Make two quarter-sheet note cards as shown. Use the cards to summarize information about ionic and metallic compounds.

WORD ORIGIN

ion
from Greek *ienai*, means "to go"

Understanding Ions

As you read in Lesson 2, the atoms of two or more nonmetals form compounds by sharing valence electrons. However, when a metal and a nonmetal bond, they do not share electrons. Instead, one or more valence electrons transfers from the metal atom to the nonmetal atom. After electrons transfer, the atoms bond and form a chemically stable compound. Transferring valence electrons results in atoms with the same number of valence electrons as a noble gas.

When an atom loses or gains a valence electron, it becomes an ion. *An **ion** is an atom that is no longer electrically neutral because it has lost or gained valence electrons.* Because electrons have a negative charge, losing or gaining an electron changes the overall charge of an atom. An atom that loses valence electrons becomes an ion with a positive charge. This is because the number of electrons is now less than the number of protons in the atom. An atom that gains valence electrons becomes an ion with a negative charge. This is because the number of protons is now less than the number of electrons.

Reading Check Why do atoms that a gain electrons become an ion with a negative charge?

Losing Valence Electrons

Look at the periodic table on the inside back cover of this book. What information about sodium (Na) can you infer from the periodic table? Sodium is a metal. Its atomic number is 11. This means each sodium atom has 11 protons and 11 electrons. Sodium is in group 1 on the periodic table. Therefore, sodium atoms have one valence electron, and they are chemically unstable.

Metal atoms, such as sodium, become more stable when they lose valence electrons and form a chemical bond with a nonmetal. If a sodium atom loses its one valence electron, it would have a total of ten electrons. Which element on the periodic table has atoms with ten electrons? Neon (Ne) atoms have a total of ten electrons. Eight of these are valence electrons. When a sodium atom loses one valence electron, the electrons in the next lower energy level are now the new valence electrons. The sodium atom then has eight valence electrons, the same as the noble gas neon and is chemically stable.

Gaining Valence Electrons

In Lesson 2, you read that nonmetal atoms can share valence electrons with other nonmetal atoms. Nonmetal atoms can also gain valence electrons from metal atoms. Either way, they achieve the electron arrangement of a noble gas. Find the nonmetal chlorine (Cl) on the periodic table. Its atomic number is 17. Atoms of chlorine have seven valence electrons. If a chlorine atom gains one valence electron, it will have eight valence electrons. It will also have the same electron arrangement as the noble gas argon (Ar).

When a sodium atom loses a valence electron, it becomes a positively charged ion. This is shown by a plus (+) sign. When a chlorine atom gains a valence electron, it becomes a negatively charged ion. This is shown by a negative (−) sign. **Figure 11** illustrates the process of a sodium atom losing an electron and a chlorine atom gaining an electron.

 Reading Check Are atoms of a group 16 element more likely to gain or lose valence electrons?

Losing and Gaining Electrons

Figure 11 Sodium atoms have a tendency to lose a valence electron. Chlorine atoms have a tendency to gain a valence electron.

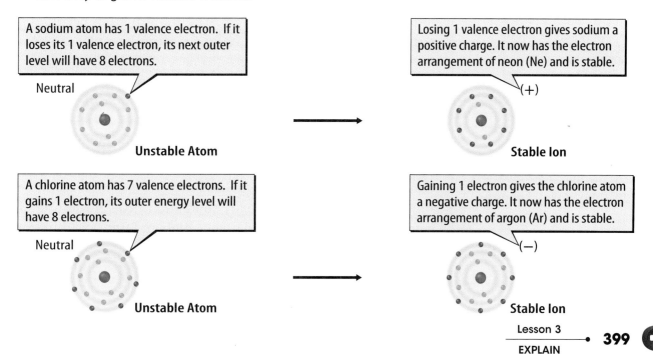

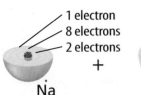
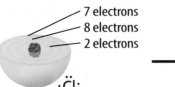
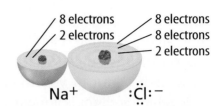

Sodium and chlorine atoms are stable when they have eight valence electrons. A sodium atoms loses one valence electron and becomes stable. A chlorine atom gains one valence electron and becomes stable.

The positively charged sodium ion and the negatively charged chlorine ion attract each other. Together they form a strong ionic bond.

Figure 12 An ionic bond forms between Na and Cl when an electron transfers from Na to Cl.

Concepts in Motion Animation

Math Skills

Use Percentage

An atom's radius is measured in picometers (pm), 1 trillion times smaller than a meter. When an atom becomes an ion, its radius increases or decreases. For example, a Na atom has a radius of **186 pm**. A Na^+ ion has a radius of **102 pm**. By what percentage does the radius change?

Subtract the atom's radius from the ion's radius.

102 pm − 186 pm = −84 pm

Divide the difference by the atom's radius.

−84 pm ÷ 186 pm = −0.45

Multiply the answer by 100 and add a % sign.

−0.45 × 100 = −45%

A negative value is a decrease in size. A positive value is an increase.

Practice

The radius of an oxygen (O) atom is 73 pm. The radius of an oxygen ion (O^{2-}) is 140 pm. By what percentage does the radius change?

- **Review**
- Math Practice
- Personal Tutor

Determining an Ion's Charge

Atoms are electrically neutral because they have the same number of protons and electrons. Once an atom gains or loses electrons, it becomes a charged ion. For example, the atomic number for nitrogen (N) is 7. Each N atom has 7 protons and 7 electrons and is electrically neutral. However, an N atom often gains 3 electrons when forming an ion. The N ion then has 10 electrons. To determine the charge, subtract the number of electrons in the ion from the number of protons.

$$7 \text{ protons} - 10 \text{ electrons} = -3 \text{ charge}$$

A nitrogen ion has a −3 charge. This is written as N^{3-}.

Ionic Bonds—Electron Transferring

Recall that metal atoms typically lose valence electrons and nonmetal atoms typically gain valence electrons. When forming a chemical bond, the nonmetal atoms gain the electrons lost by the metal atoms. Take a look at **Figure 12**. In NaCl, or table salt, a sodium atom loses a valence electron. The electron is transferred to a chlorine atom. The sodium atom becomes a positively charged ion. The chlorine atom becomes a negatively charged ion. These ions attract each other and form a stable ionic compound. *The attraction between positively and negatively charged ions in an ionic compound is an* **ionic bond.**

Key Concept Check What holds ionic compounds together?

Ionic Compounds

Ionic compounds are usually solid and brittle at room temperature. They also have relatively high melting and boiling points. Many ionic compounds dissolve in water. Water that contains dissolved ionic compounds is a good conductor of electricity. This is because an electrical charge can pass from ion to ion in the solution.

Comparing Ionic and Covalent Compounds

Recall that in a covalent bond, two or more nonmetal atoms share electrons and form a unit, or molecule. Covalent compounds, such as water, are made up of many molecules. However, when nonmetal ions bond to metal ions in an ionic compound, there are no molecules. Instead, there is a large collection of oppositely charged ions. All of the ions attract each other and are held together by ionic bonds.

Metallic Bonds— Electron Pooling

Recall that metal atoms typically lose valence electrons when forming compounds. What happens when metal atoms bond to other metal atoms? Metal atoms form compounds with one another by combining, or pooling, their valence electrons. A **metallic bond** is a bond formed when many metal atoms share their pooled valence electrons.

The pooling of valence electrons in aluminum is shown in **Figure 13**. The aluminum atoms lose their valence electrons and become positive ions, indicated by the plus (+) signs. The negative (−) signs indicate the valence electrons, which move from ion to ion. Valence electrons in metals are not bonded to one atom. Instead, a "sea of electrons" surrounds the positive ions.

 Key Concept Check How do metal atoms bond with one another?

Figure 13 Valence electrons move among all the aluminum (Al) ions.

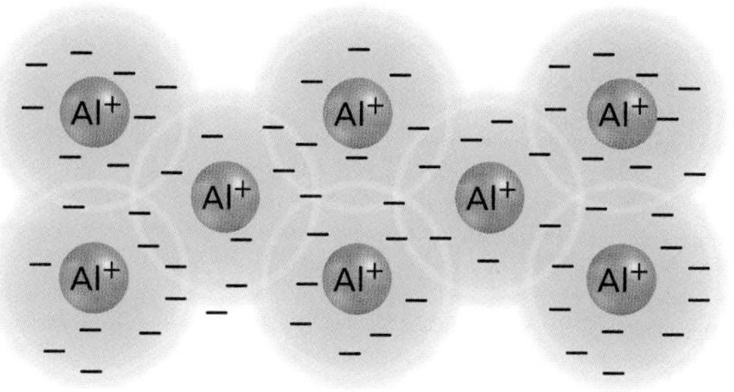

Inquiry MiniLab 20 minutes

How many ionic compounds can you make?

You have read that in ionic bonding, metal atoms transfer electrons to nonmetal atoms.

1. Copy the table below into your Science Journal.

Group	Elements	Type	Dot Diagram
1	Li, Na, K	Metal	Ẋ
2	Be, Mg, Ca	Metal	
14	C	Nonmetal	
15	N, P	Nonmetal	
16	O, S	Nonmetal	
17	F, Cl	Nonmetal	

2. Fill in the last column with the correct dot diagram for each group. Color the dots of the metal atoms with a **red marker** and the dots of the nonmetal atoms with a **blue marker.**

3. Using the information in your table, create five different ionic bonds. Write (a) the equation for the electron transfer and (b) the formula for each compound. For example:

a. $\dot{N}a + \dot{N}a + \cdot\ddot{O}: \longrightarrow Na^+ + Na^+ + :\ddot{O}:^{2-}$

b. Na_2O

Analyze and Conclude

1. **Explain** What happens to the metal and nonmetal ions after the electrons have been transferred?

2. **Key Concept** Describe the ionic bonds that hold the ions together in your compounds.

ACADEMIC VOCABULARY
conduct
(verb) to serve as a medium through which something can flow

Table 1 Bonds can form when atoms share valence electrons, transfer valence electrons, or pool valence electrons.

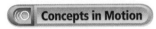
Interactive Table

Properties of Metallic Compounds

Metals are good conductors of thermal energy and electricity. Because the valence electrons can move from ion to ion, they can easily conduct an electric charge. When a metal is hammered into a sheet or drawn into a wire, it does not break. The metal ions can slide past one another in the electron sea and move to new positions. Metals are shiny because the valence electrons at the surface of a metal interact with light. **Table 1** compares the covalent, ionic, and metallic bonds that you studied in this chapter.

 Reading Check How does valence electron pooling explain why metals can be hammered into a sheet?

Table 1 Covalent, Ionic, and Metallic Bonds

Type of Bond	What is bonding?	Properties of Compounds
Covalent (Water)	nonmetal atoms; nonmetal atoms	• gas, liquid, or solid • low melting and boiling points • often not able to dissolve in water • poor conductors of thermal energy and electricity • dull appearance
Ionic (Salt)	nonmetal ions; metal ions	• solid crystals • high melting and boiling points • dissolves in water • solids are poor conductors of thermal energy and electricity • ionic compounds in water solutions conduct electricity
Metallic (Aluminum)	metal ions; metal ions	• usually solid at room temperature • high melting and boiling points • do not dissolve in water • good conductors of thermal energy and electricity • shiny surface • can be hammered into sheets and pulled into wires

402 Chapter 11
EXPLAIN

Lesson 3 Review

Visual Summary

Metal atoms lose electrons and non-metal atoms gain electrons and form stable compounds. An atom that has gained or lost an electron is an ion.

An ionic bond forms between positively and negatively charged ions.

Na⁺ :C̈l:⁻

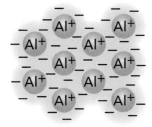

A metallic bond forms when many metal atoms share their pooled valence electrons.

FOLDABLES

Use your lesson Foldable to review the lesson. Save your Foldable for the project at the end of the chapter.

What do you think NOW?

You first read the statements below at the beginning of the chapter.

5. Losing electrons can make some atoms more chemically stable.

6. Metals are good electrical conductors because they tend to hold onto their valence electrons very tightly.

Did you change your mind about whether you agree or disagree with the statements? Rewrite any false statements to make them true.

Use Vocabulary

1. **Define** *ionic bond* in your own words.
2. An atom that changes so that it has an electrical charge is a(n) _____.
3. **Use the term** *metallic bond* in a sentence.

Understand Key Concepts

4. **Recall** What holds ionic compounds together?
5. Which element would most likely bond with lithium and form an ionic compound?
   A. beryllium   C. fluorine
   B. calcium   D. sodium
6. **Contrast** Why are metals good conductors of electricity while covalent compounds are poor conductors?

Interpret Graphics

7. **Organize** Copy and fill in the graphic organizer below. In each oval, list a common property of an ionic compound.

Critical Thinking

8. **Design** a poster to illustrate how ionic compounds form.
9. **Evaluate** What type of bonding does a material most likely have if it has a high melting point, is solid at room temperature, and easily dissolves in water?

Math Skills
— Math Practice —

10. The radius of the aluminum (Al) atom is 143 pm. The radius of the aluminum ion (Al^{3+}) is 54 pm. By what percentage did the radius change as the ion formed?

Inquiry Lab

45 minutes

Ions in Solution

Materials

250-mL beaker

plastic spoon

dull pennies (20)

stopwatch

white vinegar

table salt

iron nails (2)

sandpaper

Safety

You know that ions can combine and form stable ionic compounds. Ions can also separate in a compound and dissolve in solution. For example, pennies become dull over time because the copper ions on the surface of the pennies react with oxygen in the air and form copper(II) oxide. When you place dull pennies in a vinegar-salt solution, the copper ions separate from the oxygen ions. These ions dissolve in the solution.

Question

How do elements join together to make chemical compounds?

Procedure

1. Read and complete a lab safety form.
2. Pour 50 mL of white vinegar into a 250-mL beaker. Using a plastic spoon, add a spoonful of table salt to the vinegar. Stir the mixture with the spoon until the salt dissolves.
3. Add 20 dull pennies to the vinegar-salt solution. Leave the pennies in the solution for 10 minutes. Use a stopwatch or a clock with a second hand to measure the time.
4. After 10 minutes, use the plastic spoon to remove the pennies from the solution. Rinse the pennies in tap water. Place them on paper towels to dry. Record the change to the pennies in your Science Journal.

Form a Hypothesis

5. If you place an iron nail in the vinegar-salt solution, predict what changes will occur to the nail.

Test Your Hypothesis

6. Use sandpaper to clean two nails. Place one nail in the vinegar-salt solution, and place the other nail on a clean paper towel. You will compare the dry nail to the one in the solution and observe changes as they occur.

7. Every 5 minutes observe the nail in the solution and record your observations in your Science Journal. Remember to use the dry nail to help detect changes in the wet nail. Use a stopwatch or a clock with a second hand to measure the time. Keep the nail in the solution for 25 minutes

8. After 25 minutes, use a plastic spoon to remove the nail from the solution. Dispose of all materials as directed by your teacher.

Lab Tips

☑ Be sure the pennies are separated when they are in the vinegar-salt solution. You may need to stir them with the plastic spoon.

☑ Use the plastic spoon to bring the nail out of the solution when checking for changes.

Analyze and Conclude

9. **Compare and Contrast** What changes occurred when you placed the dull pennies in the vinegar-salt solution?

10. **Recognize Cause and Effect** What changes occurred to the nail in the leftover solution? Infer why these changes occurred.

11. **The Big Idea** Give two examples of how elements chemically combine and form compounds in this lab.

Communicate Your Results

Create a chart suitable for display summarizing this lab and your results.

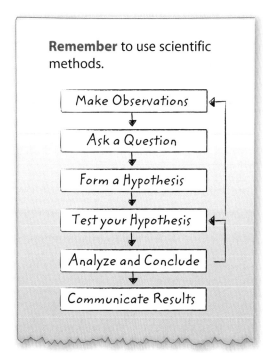

Remember to use scientific methods.

- Make Observations
- Ask a Question
- Form a Hypothesis
- Test your Hypothesis
- Analyze and Conclude
- Communicate Results

 **Extension**

The Statue of Liberty is made of copper. Research why the statue is green.

Lesson 3
EXTEND
405

Chapter 11 Study Guide

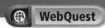

Elements can join together by sharing, transferring, or pooling electrons to make chemical compounds.

Key Concepts Summary

Lesson 1: Electrons and Energy Levels

- Electrons with more energy are farther from the atom's nucleus and are in a higher energy level.
- Atoms with fewer than eight **valence electrons** gain, lose, or share valence electrons and form stable compounds. Atoms in stable compounds have the same electron arrangement as a noble gas.

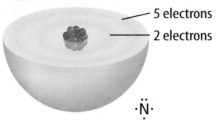

Lesson 2: Compounds, Chemical Formulas, and Covalent Bonds

- A compound and the elements it is made from have different chemical and physical properties.
- A **covalent bond** forms when two nonmetal atoms share valence electrons. Common properties of covalent compounds include low melting points and low boiling points. They are usually gas or liquid at room temperature and poor conductors of electricity.
- Water is a polar compound because the oxygen atom pulls more strongly on the shared valence electrons than the hydrogen atoms do.

Lesson 3: Ionic and Metallic Bonds

- **Ionic bonds** form when valence electrons move from a metal atom to a nonmetal atom.
- An ionic compound is held together by ionic bonds, which are attractions between positively and negatively charged **ions.**
- A **metallic bond** forms when valence electrons are pooled among many metal atoms.

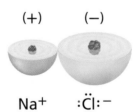

Vocabulary

chemical bond p. 382
valence electron p. 384
electron dot diagram p. 385

covalent bond p. 391
molecule p. 392
polar molecule p. 393
chemical formula p. 394

ion p. 398
ionic bond p. 400
metallic bond p. 401

Study Guide

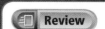

- Personal Tutor
- Vocabulary eGames
- Vocabulary eFlashcards

FOLDABLES Chapter Project

Assemble your lesson Foldables as shown to make a Chapter Project. Use the project to review what you have learned in this chapter.

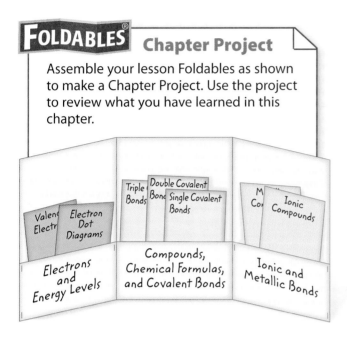

Use Vocabulary

1. The force that holds atoms together is called a(n) _____.

2. You can predict the number of bonds an atom can form by drawing its _____.

3. The nitrogen and hydrogen atoms that make up ammonia (NH_3) are held together by a(n) _____ because the atoms share valence electrons unequally.

4. Two hydrogen atoms and one oxygen atom together are a _____ of water.

5. A positively charged sodium ion and a negatively charged chlorine ion are joined by a(n) _____ to form the compound sodium chloride.

Link Vocabulary and Key Concepts

Interactive Concept Map

Copy this concept map, and then use vocabulary terms from the previous page and other terms from the chapter to complete the concept map.

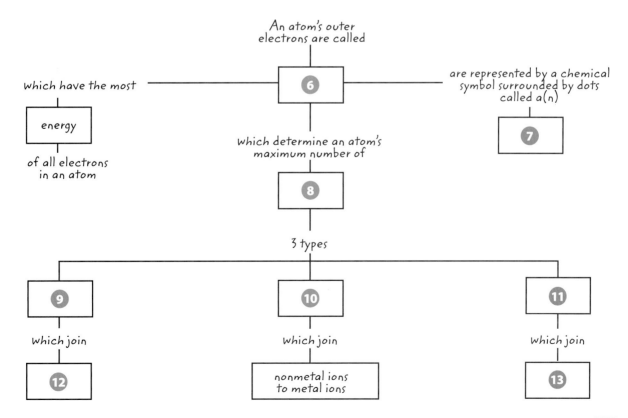

Chapter 11 Study Guide • 407

Chapter 11 Review

Understand Key Concepts

1. Atoms lose, gain, or share electrons and become as chemically stable as
   A. an electron.
   B. an ion.
   C. a metal.
   D. a noble gas.

2. Which is the correct electron dot diagram for boron, one of the group 13 elements?
   A. ∙B∙ (with dot above)
   B. ∙B: (with dots above)
   C. :B: (with dots above)
   D. ∙B∙ (with dot below)

3. If an electron transfers from one atom to another atom, what type of bond will most likely form?
   A. covalent
   B. ionic
   C. metallic
   D. polar

4. What change would make an atom represented by this diagram have the same electron arrangement as a noble gas?

   A. gaining two electrons
   B. gaining four electrons
   C. losing two electrons
   D. losing four electrons

5. What would make bromine, a group 17 element, more similar to a noble gas?
   A. gaining one electron
   B. gaining two electrons
   C. losing one electron
   D. losing two electrons

6. Which would most likely be joined by an ionic bond?
   A. a positive metal ion and a positive nonmetal ion
   B. a positive metal ion and a negative nonmetal ion
   C. a negative metal ion and a positive nonmetal ion
   D. a negative metal ion and a negative nonmetal ion

7. Which group of elements on the periodic table forms covalent compounds with other nonmetals?
   A. group 1
   B. group 2
   C. group 17
   D. group 18

8. Which best describes an atom represented by this diagram?

   A. It is likely to bond by gaining six electrons.
   B. It is likely to bond by losing two electrons.
   C. It is not likely to bond because it is already stable.
   D. It is not likely to bond because it has too few electrons.

9. How many dots would a dot diagram for selenium, one of the group 16 elements, have?
   A. 6
   B. 8
   C. 10
   D. 16

Chapter Review

Critical Thinking

10. Classify Use the periodic table to classify the elements potassium (K), bromine (Br), and argon (Ar) according to how likely their atoms are to do the following.
  a. lose electrons to form positive ions
  b. gain electrons to form negative ions
  c. neither gain nor lose electrons

11. Describe the change that is shown in this illustration. How does this change affect the stability of the atom?

$$\cdot \ddot{N} \cdot \longrightarrow :\ddot{N}:^{3-}$$

12. Analyze One of your classmates draws an electron dot diagram for a helium atom with two dots. He tells you that these dots mean each helium atom has two unpaired electrons and can gain, lose, or share electrons to have four pairs of valence electrons and become stable. What is wrong with your classmate's argument?

13. Explain why the hydrogen atoms in a hydrogen gas molecule (H_2) form nonpolar covalent bonds but the oxygen and hydrogen atoms in water molecules (H_2O) form polar covalent bonds.

14. Contrast Why is it possible for an oxygen atom to form a double covalent bond, but it is not possible for a chlorine atom to form a double covalent bond?

Writing in Science

15. Compose a poem at least ten lines long that explains ionic bonding, covalent bonding, and metallic bonding.

REVIEW THE BIG IDEA

16. Which types of atoms pool their valence electrons to form a "sea of electrons"?

17. Describe a way in which elements joining together to form chemical compounds is similar to the way the letters on a computer keyboard join together to form words.

Math Skills

Math Practice

Element	Atomic Radius	Ionic Radius
Potassium (K)	227 pm	133 pm
Iodine (I)	133 pm	216 pm

18. What is the percent change when an iodine atom (I) becomes an ion (I^-)?

19. What is the percent change when a potassium atom (K) becomes an ion (K^+)?

Standardized Test Practice

Record your answers on the answer sheet provided by your teacher or on a sheet of paper.

Multiple Choice

1. Which information does the chemical formula CO_2 NOT give you?
   A number of valence electrons in each atom
   B ratio of atoms in the compound
   C total number of atoms in one molecule of the compound
   D type of elements in the compound

Use the diagram below to answer question 2.

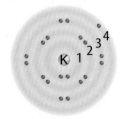

2. The diagram above shows a potassium atom. Which is the second-highest energy level?
   A 1
   B 2
   C 3
   D 4

3. What is shared in a metallic bond?
   A negatively charged ions
   B neutrons
   C pooled valence electrons
   D protons

4. Which is a characteristic of most nonpolar compounds?
   A conduct electricity poorly
   B dissolve easily in water
   C solid crystals
   D shiny surfaces

Use the diagram below to answer question 5.

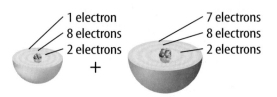

5. The atoms in the diagram above are forming a bond. Which represents that bond?

   A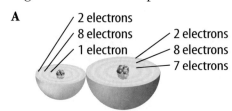

   B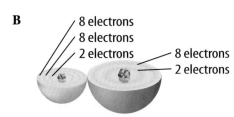

   C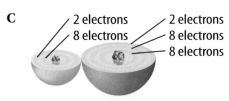

   D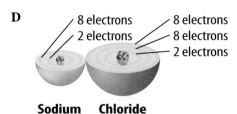

6. Covalent bonds typically form between the atoms of elements that share
   A nuclei.
   B oppositely charged ions.
   C protons.
   D valence electrons.

410 • Chapter 11 Standardized Test Practice

Standardized Test Practice

Use the diagram below to answer question 7.

Water Molecule

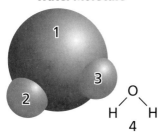

7 In the diagram above, which shows an atom with a partial negative charge?

  A 1
  B 2
  C 3
  D 4

8 Which compound is formed by the attraction between negatively and positively charged ions?

  A bipolar
  B covalent
  C ionic
  D nonpolar

9 The atoms of noble gases do NOT bond easily with other atoms because their valence electrons are

  A absent.
  B moving.
  C neutral.
  D stable.

Constructed Response

Use the table below to answer question 10.

Property	Rust	Iron	Oxygen
Color			Clear
Solid, liquid, or gas			
Strength		Strong	Does NOT apply
Usefulness			

10 Rust is a compound of iron and oxygen. Compare the properties of rust, iron, and oxygen by filling in the missing cells in the table above. What can you conclude about the properties of compounds and their elements?

Use the diagram below to answer questions 11 and 12.

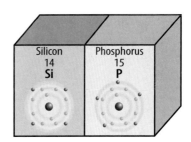

11 In the diagram, how are valence electrons illustrated? How many valence electrons does each element have?

12 Describe a stable electron configuration. For each element above, how many electrons are needed to make a stable electron configuration?

NEED EXTRA HELP?												
If You Missed Question...	1	2	3	4	5	6	7	8	9	10	11	12
Go to Lesson...	2	1	3	3	3	2	2	3	1	2	1	1

Student Resources

For Students and Parents/Guardians

These resources are designed to help you achieve success in science. You will find useful information on laboratory safety, math skills, and science skills. In addition, science reference materials are found in the Reference Handbook. You'll find the information you need to learn and sharpen your skills in these resources.

Table of Contents

Science Skill Handbook SR-2
Scientific Methods SR-2
  Identify a Question SR-2
  Gather and Organize Information SR-2
  Form a Hypothesis SR-5
  Test the Hypothesis SR-6
  Collect Data SR-6
  Analyze the Data SR-9
  Draw Conclustions SR-10
  Communicate SR-10
Safety Symbols SR-11
Safety in the Science Laboratory SR-12
  General Safety Rules SR-12
  Prevent Accidents SR-12
  Laboratory Work SR-13
  Emergencies SR-13

Math Skill Handbook SR-14
Math Review SR-14
  Use Fractions SR-14
  Use Ratios SR-17
  Use Decimals SR-17
  Use Proportions SR-18
  Use Percentages SR-19
  Solve One-Step Equations SR-19
  Use Statistics SR-20
  Use Geometry SR-21
Science Application SR-24
  Measure in SI SR-24
  Dimensional Analysis SR-24
  Precision and Significant Digits SR-26
  Scientific Notation SR-26
  Make and Use Graphs SR-27

Foldables Handbook SR-29

Reference Handbook SR-40
  Periodic Table of the Elements SR-40

Glossary G-2

Index I-2

Credits C-2

Science Skill Handbook

Scientific Methods

Scientists use an orderly approach called the scientific method to solve problems. This includes organizing and recording data so others can understand them. Scientists use many variations in this method when they solve problems.

Identify a Question

The first step in a scientific investigation or experiment is to identify a question to be answered or a problem to be solved. For example, you might ask which gasoline is the most efficient.

Gather and Organize Information

After you have identified your question, begin gathering and organizing information. There are many ways to gather information, such as researching in a library, interviewing those knowledgeable about the subject, and testing and working in the laboratory and field. Fieldwork is investigations and observations done outside of a laboratory.

Researching Information Before moving in a new direction, it is important to gather the information that already is known about the subject. Start by asking yourself questions to determine exactly what you need to know. Then you will look for the information in various reference sources, like the student is doing in **Figure 1.** Some sources may include textbooks, encyclopedias, government documents, professional journals, science magazines, and the Internet. Always list the sources of your information.

Figure 1 The Internet can be a valuable research tool.

Evaluate Sources of Information Not all sources of information are reliable. You should evaluate all of your sources of information, and use only those you know to be dependable. For example, if you are researching ways to make homes more energy efficient, a site written by the U.S. Department of Energy would be more reliable than a site written by a company that is trying to sell a new type of weatherproofing material. Also, remember that research always is changing. Consult the most current resources available to you. For example, a 1985 resource about saving energy would not reflect the most recent findings.

Sometimes scientists use data that they did not collect themselves, or conclusions drawn by other researchers. This data must be evaluated carefully. Ask questions about how the data were obtained, if the investigation was carried out properly, and if it has been duplicated exactly with the same results. Would you reach the same conclusion from the data? Only when you have confidence in the data can you believe it is true and feel comfortable using it.

Interpret Scientific Illustrations As you research a topic in science, you will see drawings, diagrams, and photographs to help you understand what you read. Some illustrations are included to help you understand an idea that you can't see easily by yourself, like the tiny particles in an atom in **Figure 2**. A drawing helps many people to remember details more easily and provides examples that clarify difficult concepts or give additional information about the topic you are studying. Most illustrations have labels or a caption to identify or to provide more information.

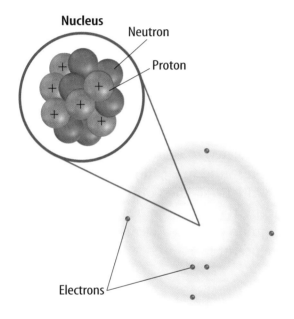

Figure 2 This drawing shows an atom of carbon with its six protons, six neutrons, and six electrons.

Concept Maps One way to organize data is to draw a diagram that shows relationships among ideas (or concepts). A concept map can help make the meanings of ideas and terms more clear, and help you understand and remember what you are studying. Concept maps are useful for breaking large concepts down into smaller parts, making learning easier.

Network Tree A type of concept map that not only shows a relationship, but how the concepts are related is a network tree, shown in **Figure 3**. In a network tree, the words are written in the ovals, while the description of the type of relationship is written across the connecting lines.

When constructing a network tree, write down the topic and all major topics on separate pieces of paper or notecards. Then arrange them in order from general to specific. Branch the related concepts from the major concept and describe the relationship on the connecting line. Continue to more specific concepts until finished.

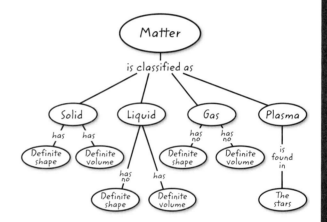

Figure 3 A network tree shows how concepts or objects are related.

Events Chain Another type of concept map is an events chain. Sometimes called a flow chart, it models the order or sequence of items. An events chain can be used to describe a sequence of events, the steps in a procedure, or the stages of a process.

When making an events chain, first find the one event that starts the chain. This event is called the initiating event. Then, find the next event and continue until the outcome is reached, as shown in **Figure 4** on the next page.

Science Skill Handbook • **SR-3**

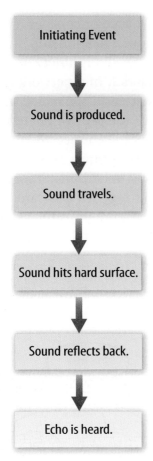

Figure 4 Events-chain concept maps show the order of steps in a process or event. This concept map shows how a sound makes an echo.

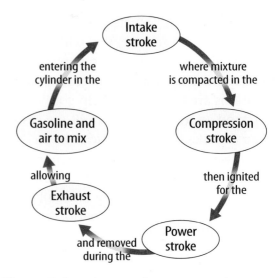

Figure 5 A cycle map shows events that occur in a cycle.

Cycle Map A specific type of events chain is a cycle map. It is used when the series of events do not produce a final outcome, but instead relate back to the beginning event, such as in **Figure 5**. Therefore, the cycle repeats itself.

To make a cycle map, first decide what event is the beginning event. This is also called the initiating event. Then list the next events in the order that they occur, with the last event relating back to the initiating event. Words can be written between the events that describe what happens from one event to the next. The number of events in a cycle map can vary, but usually contain three or more events.

Spider Map A type of concept map that you can use for brainstorming is the spider map. When you have a central idea, you might find that you have a jumble of ideas that relate to it but are not necessarily clearly related to each other. The spider map on sound in **Figure 6** shows that if you write these ideas outside the main concept, then you can begin to separate and group unrelated terms so they become more useful.

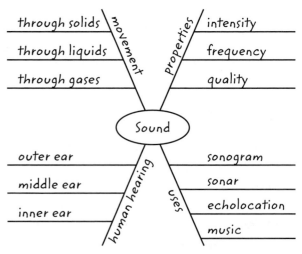

Figure 6 A spider map allows you to list ideas that relate to a central topic but not necessarily to one another.

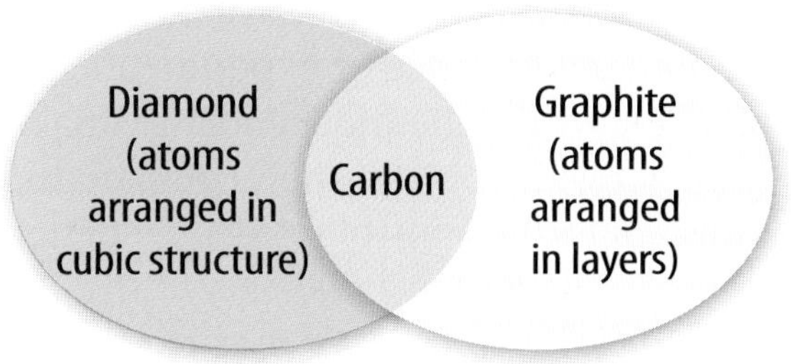

Figure 7 This Venn diagram compares and contrasts two substances made from carbon.

Venn Diagram To illustrate how two subjects compare and contrast you can use a Venn diagram. You can see the characteristics that the subjects have in common and those that they do not, shown in **Figure 7**.

To create a Venn diagram, draw two overlapping ovals that are big enough to write in. List the characteristics unique to one subject in one oval, and the characteristics of the other subject in the other oval. The characteristics in common are listed in the overlapping section.

Make and Use Tables One way to organize information so it is easier to understand is to use a table. Tables can contain numbers, words, or both.

To make a table, list the items to be compared in the first column and the characteristics to be compared in the first row. The title should clearly indicate the content of the table, and the column or row heads should be clear. Notice that in **Table 1** the units are included.

Table 1 Recyclables Collected During Week			
Day of Week	Paper (kg)	Aluminum (kg)	Glass (kg)
Monday	5.0	4.0	12.0
Wednesday	4.0	1.0	10.0
Friday	2.5	2.0	10.0

Make a Model One way to help you better understand the parts of a structure, the way a process works, or to show things too large or small for viewing is to make a model. For example, an atomic model made of a plastic-ball nucleus and chenille stem electron shells can help you visualize how the parts of an atom relate to each other. Other types of models can be devised on a computer or represented by equations.

Form a Hypothesis

A possible explanation based on previous knowledge and observations is called a hypothesis. After researching gasoline types and recalling previous experiences in your family's car you form a hypothesis—our car runs more efficiently because we use premium gasoline. To be valid, a hypothesis has to be something you can test by using an investigation.

Predict When you apply a hypothesis to a specific situation, you predict something about that situation. A prediction makes a statement in advance, based on prior observation, experience, or scientific reasoning. People use predictions to make everyday decisions. Scientists test predictions by performing investigations. Based on previous observations and experiences, you might form a prediction that cars are more efficient with premium gasoline. The prediction can be tested in an investigation.

Design an Experiment A scientist needs to make many decisions before beginning an investigation. Some of these include: how to carry out the investigation, what steps to follow, how to record the data, and how the investigation will answer the question. It also is important to address any safety concerns.

Test the Hypothesis

Now that you have formed your hypothesis, you need to test it. Using an investigation, you will make observations and collect data, or information. This data might either support or not support your hypothesis. Scientists collect and organize data as numbers and descriptions.

Follow a Procedure In order to know what materials to use, as well as how and in what order to use them, you must follow a procedure. **Figure 8** shows a procedure you might follow to test your hypothesis.

Procedure

Step 1	Use regular gasoline for two weeks.
Step 2	Record the number of kilometers between fill-ups and the amount of gasoline used.
Step 3	Switch to premium gasoline for two weeks.
Step 4	Record the number of kilometers between fill-ups and the amount of gasoline used.

Figure 8 A procedure tells you what to do step-by-step.

Identify and Manipulate Variables and Controls In any experiment, it is important to keep everything the same except for the item you are testing. The one factor you change is called the independent variable. The change that results is the dependent variable. Make sure you have only one independent variable, to assure yourself of the cause of the changes you observe in the dependent variable. For example, in your gasoline experiment the type of fuel is the independent variable. The dependent variable is the efficiency.

Many experiments also have a control—an individual instance or experimental subject for which the independent variable is not changed. You can then compare the test results to the control results. To design a control you can have two cars of the same type. The control car uses regular gasoline for four weeks. After you are done with the test, you can compare the experimental results to the control results.

Collect Data

Whether you are carrying out an investigation or a short observational experiment, you will collect data, as shown in **Figure 9.** Scientists collect data as numbers and descriptions and organize them in specific ways.

Observe Scientists observe items and events, then record what they see. When they use only words to describe an observation, it is called qualitative data. Scientists' observations also can describe how much there is of something. These observations use numbers, as well as words, in the description and are called quantitative data. For example, if a sample of the element gold is described as being "shiny and very dense" the data are qualitative. Quantitative data on this sample of gold might include "a mass of 30 g and a density of 19.3 g/cm^3."

Figure 9 Collecting data is one way to gather information directly.

Figure 10 Record data neatly and clearly so it is easy to understand.

When you make observations you should examine the entire object or situation first, and then look carefully for details. It is important to record observations accurately and completely. Always record your notes immediately as you make them, so you do not miss details or make a mistake when recording results from memory. Never put unidentified observations on scraps of paper. Instead they should be recorded in a notebook, like the one in **Figure 10.** Write your data neatly so you can easily read it later. At each point in the experiment, record your observations and label them. That way, you will not have to determine what the figures mean when you look at your notes later. Set up any tables that you will need to use ahead of time, so you can record any observations right away. Remember to avoid bias when collecting data by not including personal thoughts when you record observations. Record only what you observe.

Estimate Scientific work also involves estimating. To estimate is to make a judgment about the size or the number of something without measuring or counting. This is important when the number or size of an object or population is too large or too difficult to accurately count or measure.

Sample Scientists may use a sample or a portion of the total number as a type of estimation. To sample is to take a small, representative portion of the objects or organisms of a population for research. By making careful observations or manipulating variables within that portion of the group, information is discovered and conclusions are drawn that might apply to the whole population. A poorly chosen sample can be unrepresentative of the whole. If you were trying to determine the rainfall in an area, it would not be best to take a rainfall sample from under a tree.

Measure You use measurements every day. Scientists also take measurements when collecting data. When taking measurements, it is important to know how to use measuring tools properly. Accuracy also is important.

Length To measure length, the distance between two points, scientists use meters. Smaller measurements might be measured in centimeters or millimeters.

Length is measured using a metric ruler or meterstick. When using a metric ruler, line up the 0-cm mark with the end of the object being measured and read the number of the unit where the object ends. Look at the metric ruler shown in **Figure 11.** The centimeter lines are the long, numbered lines, and the shorter lines are millimeter lines. In this instance, the length would be 4.50 cm.

Figure 11 This metric ruler has centimeter and millimeter divisions.

Science Skill Handbook • **SR-7**

Mass The SI unit for mass is the kilogram (kg). Scientists can measure mass using units formed by adding metric prefixes to the unit gram (g), such as milligram (mg). To measure mass, you might use a triple-beam balance similar to the one shown in **Figure 12**. The balance has a pan on one side and a set of beams on the other side. Each beam has a rider that slides on the beam.

When using a triple-beam balance, place an object on the pan. Slide the largest rider along its beam until the pointer drops below zero. Then move it back one notch. Repeat the process for each rider proceeding from the larger to smaller until the pointer swings an equal distance above and below the zero point. Sum the masses on each beam to find the mass of the object. Move all riders back to zero when finished.

Instead of putting materials directly on the balance, scientists often take a tare of a container. A tare is the mass of a container into which objects or substances are placed for measuring their masses. To find the mass of objects or substances, find the mass of a clean container. Remove the container from the pan, and place the object or substances in the container. Find the mass of the container with the materials in it. Subtract the mass of the empty container from the mass of the filled container to find the mass of the materials you are using.

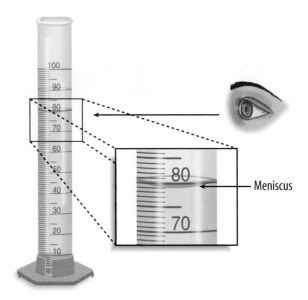

Figure 13 Graduated cylinders measure liquid volume.

Liquid Volume To measure liquids, the unit used is the liter. When a smaller unit is needed, scientists might use a milliliter. Because a milliliter takes up the volume of a cube measuring 1 cm on each side it also can be called a cubic centimeter ($cm^3 = cm \times cm \times cm$).

You can use beakers and graduated cylinders to measure liquid volume. A graduated cylinder, shown in **Figure 13**, is marked from bottom to top in milliliters. In lab, you might use a 10-mL graduated cylinder or a 100-mL graduated cylinder. When measuring liquids, notice that the liquid has a curved surface. Look at the surface at eye level, and measure the bottom of the curve. This is called the meniscus. The graduated cylinder in **Figure 13** contains 79.0 mL, or 79.0 cm^3, of a liquid.

Temperature Scientists often measure temperature using the Celsius scale. Pure water has a freezing point of 0°C and boiling point of 100°C. The unit of measurement is degrees Celsius. Two other scales often used are the Fahrenheit and Kelvin scales.

Figure 12 A triple-beam balance is used to determine the mass of an object.

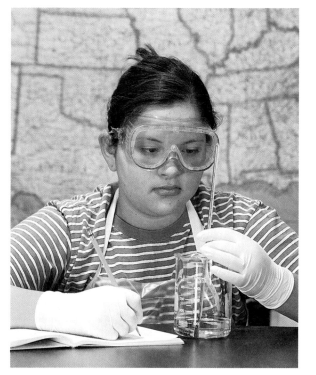

Figure 14 A thermometer measures the temperature of an object.

Scientists use a thermometer to measure temperature. Most thermometers in a laboratory are glass tubes with a bulb at the bottom end containing a liquid such as colored alcohol. The liquid rises or falls with a change in temperature. To read a glass thermometer like the thermometer in **Figure 14,** rotate it slowly until a red line appears. Read the temperature where the red line ends.

Form Operational Definitions An operational definition defines an object by how it functions, works, or behaves. For example, when you are playing hide and seek and a tree is home base, you have created an operational definition for a tree.

Objects can have more than one operational definition. For example, a ruler can be defined as a tool that measures the length of an object (how it is used). It can also be a tool with a series of marks used as a standard when measuring (how it works).

Analyze the Data

To determine the meaning of your observations and investigation results, you will need to look for patterns in the data. Then you must think critically to determine what the data mean. Scientists use several approaches when they analyze the data they have collected and recorded. Each approach is useful for identifying specific patterns.

Interpret Data The word *interpret* means "to explain the meaning of something." When analyzing data from an experiment, try to find out what the data show. Identify the control group and the test group to see whether changes in the independent variable have had an effect. Look for differences in the dependent variable between the control and test groups.

Classify Sorting objects or events into groups based on common features is called classifying. When classifying, first observe the objects or events to be classified. Then select one feature that is shared by some members in the group, but not by all. Place those members that share that feature in a subgroup. You can classify members into smaller and smaller subgroups based on characteristics. Remember that when you classify, you are grouping objects or events for a purpose. Keep your purpose in mind as you select the features to form groups and subgroups.

Compare and Contrast Observations can be analyzed by noting the similarities and differences between two or more objects or events that you observe. When you look at objects or events to see how they are similar, you are comparing them. Contrasting is looking for differences in objects or events.

Science Skill Handbook • **SR-9**

Recognize Cause and Effect A cause is a reason for an action or condition. The effect is that action or condition. When two events happen together, it is not necessarily true that one event caused the other. Scientists must design a controlled investigation to recognize the exact cause and effect.

Draw Conclusions

When scientists have analyzed the data they collected, they proceed to draw conclusions about the data. These conclusions are sometimes stated in words similar to the hypothesis that you formed earlier. They may confirm a hypothesis, or lead you to a new hypothesis.

Infer Scientists often make inferences based on their observations. An inference is an attempt to explain observations or to indicate a cause. An inference is not a fact, but a logical conclusion that needs further investigation. For example, you may infer that a fire has caused smoke. Until you investigate, however, you do not know for sure.

Apply When you draw a conclusion, you must apply those conclusions to determine whether the data supports the hypothesis. If your data do not support your hypothesis, it does not mean that the hypothesis is wrong. It means only that the result of the investigation did not support the hypothesis. Maybe the experiment needs to be redesigned, or some of the initial observations on which the hypothesis was based were incomplete or biased. Perhaps more observation or research is needed to refine your hypothesis. A successful investigation does not always come out the way you originally predicted.

Avoid Bias Sometimes a scientific investigation involves making judgments. When you make a judgment, you form an opinion. It is important to be honest and not to allow any expectations of results to bias your judgments. This is important throughout the entire investigation, from researching to collecting data to drawing conclusions.

Communicate

The communication of ideas is an important part of the work of scientists. A discovery that is not reported will not advance the scientific community's understanding or knowledge. Communication among scientists also is important as a way of improving their investigations.

Scientists communicate in many ways, from writing articles in journals and magazines that explain their investigations and experiments, to announcing important discoveries on television and radio. Scientists also share ideas with colleagues on the Internet or present them as lectures, like the student is doing in **Figure 15**.

Figure 15 A student communicates to his peers about his investigation.

These safety symbols are used in laboratory and field investigations in this book to indicate possible hazards. Learn the meaning of each symbol and refer to this page often. *Remember to wash your hands thoroughly after completing lab procedures.*

PROTECTIVE EQUIPMENT   Do not begin any lab without the proper protection equipment.

 GOGGLES Proper eye protection must be worn when performing or observing science activities that involve items or conditions as listed below.

 APRON Wear an approved apron when using substances that could stain, wet, or destroy cloth.

 SOAP Wash hands with soap and water before removing goggles and after all lab activities.

 GLOVES Wear gloves when working with biological materials, chemicals, animals, or materials that can stain or irritate hands.

LABORATORY HAZARDS

Symbols	Potential Hazards	Precaution	Response
DISPOSAL	contamination of classroom or environment due to improper disposal of materials such as chemicals and live specimens	• DO NOT dispose of hazardous materials in the sink or trash can. • Dispose of wastes as directed by your teacher.	• If hazardous materials are disposed of improperly, notify your teacher immediately.
EXTREME TEMPERATURE	skin burns due to extremely hot or cold materials such as hot glass, liquids, or metals; liquid nitrogen; dry ice	• Use proper protective equipment, such as hot mitts and/or tongs, when handling objects with extreme temperatures.	• If injury occurs, notify your teacher immediately.
SHARP OBJECTS	punctures or cuts from sharp objects such as razor blades, pins, scalpels, and broken glass	• Handle glassware carefully to avoid breakage. • Walk with sharp objects pointed downward, away from you and others.	• If broken glass or injury occurs, notify your teacher immediately.
ELECTRICAL	electric shock or skin burn due to improper grounding, short circuits, liquid spills, or exposed wires	• Check condition of wires and apparatus for fraying or uninsulated wires, and broken or cracked equipment. • Use only GFCI-protected outlets	• DO NOT attempt to fix electrical problems. Notify your teacher immediately.
CHEMICAL	skin irritation or burns, breathing difficulty, and/or poisoning due to touching, swallowing, or inhalation of chemicals such as acids, bases, bleach, metal compounds, iodine, poinsettias, pollen, ammonia, acetone, nail polish remover, heated chemicals, mothballs, and any other chemicals labeled or known to be dangerous	• Wear proper protective equipment such as goggles, apron, and gloves when using chemicals. • Ensure proper room ventilation or use a fume hood when using materials that produce fumes. • NEVER smell fumes directly. • NEVER taste or eat any material in the laboratory.	• If contact occurs, immediately flush affected area with water and notify your teacher. • If a spill occurs, leave the area immediately and notify your teacher.
FLAMMABLE	unexpected fire due to liquids or gases that ignite easily such as rubbing alcohol	• Avoid open flames, sparks, or heat when flammable liquids are present.	• If a fire occurs, leave the area immediately and notify your teacher.
OPEN FLAME	burns or fire due to open flame from matches, Bunsen burners, or burning materials	• Tie back loose hair and clothing. • Keep flame away from all materials. • Follow teacher instructions when lighting and extinguishing flames. • Use proper protection, such as hot mitts or tongs, when handling hot objects.	• If a fire occurs, leave the area immediately and notify your teacher.
ANIMAL SAFETY	injury to or from laboratory animals	• Wear proper protective equipment such as gloves, apron, and goggles when working with animals. • Wash hands after handling animals.	• If injury occurs, notify your teacher immediately.
BIOLOGICAL	infection or adverse reaction due to contact with organisms such as bacteria, fungi, and biological materials such as blood, animal or plant materials	• Wear proper protective equipment such as gloves, goggles, and apron when working with biological materials. • Avoid skin contact with an organism or any part of the organism. • Wash hands after handling organisms.	• If contact occurs, wash the affected area and notify your teacher immediately.
FUME	breathing difficulties from inhalation of fumes from substances such as ammonia, acetone, nail polish remover, heated chemicals, and mothballs	• Wear goggles, apron, and gloves. • Ensure proper room ventilation or use a fume hood when using substances that produce fumes. • NEVER smell fumes directly.	• If a spill occurs, leave area and notify your teacher immediately.
IRRITANT	irritation of skin, mucous membranes, or respiratory tract due to materials such as acids, bases, bleach, pollen, mothballs, steel wool, and potassium permanganate	• Wear goggles, apron, and gloves. • Wear a dust mask to protect against fine particles.	• If skin contact occurs, immediately flush the affected area with water and notify your teacher.
RADIOACTIVE	excessive exposure from alpha, beta, and gamma particles	• Remove gloves and wash hands with soap and water before removing remainder of protective equipment.	• If cracks or holes are found in the container, notify your teacher immediately.

Safety in the Science Laboratory

Introduction to Science Safety

The science laboratory is a safe place to work if you follow standard safety procedures. Being responsible for your own safety helps to make the entire laboratory a safer place for everyone. When performing any lab, read and apply the caution statements and safety symbol listed at the beginning of the lab.

General Safety Rules

1. Complete the *Lab Safety Form* or other safety contract BEFORE starting any science lab.
2. Study the procedure. Ask your teacher any questions. Be sure you understand safety symbols shown on the page.
3. Notify your teacher about allergies or other health conditions that can affect your participation in a lab.
4. Learn and follow use and safety procedures for your equipment. If unsure, ask your teacher.

5. Never eat, drink, chew gum, apply cosmetics, or do any personal grooming in the lab. Never use lab glassware as food or drink containers. Keep your hands away from your face and mouth.
6. Know the location and proper use of the safety shower, eye wash, fire blanket, and fire alarm.

Prevent Accidents

1. Use the safety equipment provided to you. Goggles and a safety apron should be worn during investigations.
2. Do NOT use hair spray, mousse, or other flammable hair products. Tie back long hair and tie down loose clothing.
3. Do NOT wear sandals or other open-toed shoes in the lab.
4. Remove jewelry on hands and wrists. Loose jewelry, such as chains and long necklaces, should be removed to prevent them from getting caught in equipment.
5. Do not taste any substances or draw any material into a tube with your mouth.
6. Proper behavior is expected in the lab. Practical jokes and fooling around can lead to accidents and injury.
7. Keep your work area uncluttered.

Laboratory Work

1. Collect and carry all equipment and materials to your work area before beginning a lab.
2. Remain in your own work area unless given permission by your teacher to leave it.

3. Always slant test tubes away from yourself and others when heating them, adding substances to them, or rinsing them.
4. If instructed to smell a substance in a container, hold the container a short distance away and fan vapors toward your nose.
5. Do NOT substitute other chemicals/substances for those in the materials list unless instructed to do so by your teacher.
6. Do NOT take any materials or chemicals outside of the laboratory.
7. Stay out of storage areas unless instructed to be there and supervised by your teacher.

Laboratory Cleanup

1. Turn off all burners, water, and gas, and disconnect all electrical devices.
2. Clean all pieces of equipment and return all materials to their proper places.
3. Dispose of chemicals and other materials as directed by your teacher. Place broken glass and solid substances in the proper containers. Never discard materials in the sink.
4. Clean your work area.
5. Wash your hands with soap and water thoroughly BEFORE removing your goggles.

Emergencies

1. Report any fire, electrical shock, glassware breakage, spill, or injury, no matter how small, to your teacher immediately. Follow his or her instructions.
2. If your clothing should catch fire, STOP, DROP, and ROLL. If possible, smother it with the fire blanket or get under a safety shower. NEVER RUN.
3. If a fire should occur, turn off all gas and leave the room according to established procedures.
4. In most instances, your teacher will clean up spills. Do NOT attempt to clean up spills unless you are given permission and instructions to do so.
5. If chemicals come into contact with your eyes or skin, notify your teacher immediately. Use the eyewash, or flush your skin or eyes with large quantities of water.
6. The fire extinguisher and first-aid kit should only be used by your teacher unless it is an extreme emergency and you have been given permission.
7. If someone is injured or becomes ill, only a professional medical provider or someone certified in first aid should perform first-aid procedures.

Science Skill Handbook • **SR-13**

Math Skill Handbook

Math Review

Use Fractions

A fraction compares a part to a whole. In the fraction $\frac{2}{3}$, the 2 represents the part and is the numerator. The 3 represents the whole and is the denominator.

Reduce Fractions To reduce a fraction, you must find the largest factor that is common to both the numerator and the denominator, the greatest common factor (GCF). Divide both numbers by the GCF. The fraction has then been reduced, or it is in its simplest form.

Example

Twelve of the 20 chemicals in the science lab are in powder form. What fraction of the chemicals used in the lab are in powder form?

Step 1 Write the fraction.

$$\frac{part}{whole} = \frac{12}{20}$$

Step 2 To find the GCF of the numerator and denominator, list all of the factors of each number.

Factors of 12: 1, 2, 3, 4, 6, 12 (the numbers that divide evenly into 12)

Factors of 20: 1, 2, 4, 5, 10, 20 (the numbers that divide evenly into 20)

Step 3 List the common factors.

1, 2, 4

Step 4 Choose the greatest factor in the list. The GCF of 12 and 20 is 4.

Step 5 Divide the numerator and denominator by the GCF.

$$\frac{12 \div 4}{20 \div 4} = \frac{3}{5}$$

In the lab, $\frac{3}{5}$ of the chemicals are in powder form.

Practice Problem At an amusement park, 66 of 90 rides have a height restriction. What fraction of the rides, in its simplest form, has a height restriction?

Add and Subtract Fractions with Like Denominators To add or subtract fractions with the same denominator, add or subtract the numerators and write the sum or difference over the denominator. After finding the sum or difference, find the simplest form for your fraction.

Example 1

In the forest outside your house, $\frac{1}{8}$ of the animals are rabbits, $\frac{3}{8}$ are squirrels, and the remainder are birds and insects. How many are mammals?

Step 1 Add the numerators.

$$\frac{1}{8} + \frac{3}{8} = \frac{(1+3)}{8} = \frac{4}{8}$$

Step 2 Find the GCF.

$\frac{4}{8}$ (GCF, 4)

Step 3 Divide the numerator and denominator by the GCF.

$$\frac{4 \div 4}{8 \div 4} = \frac{1}{2}$$

$\frac{1}{2}$ of the animals are mammals.

Example 2

If $\frac{7}{16}$ of the Earth is covered by freshwater, and $\frac{1}{16}$ of that is in glaciers, how much freshwater is not frozen?

Step 1 Subtract the numerators.

$$\frac{7}{16} - \frac{1}{16} = \frac{(7-1)}{16} = \frac{6}{16}$$

Step 2 Find the GCF.

$\frac{6}{16}$ (GCF, 2)

Step 3 Divide the numerator and denominator by the GCF.

$$\frac{6 \div 2}{16 \div 2} = \frac{3}{8}$$

$\frac{3}{8}$ of the freshwater is not frozen.

Practice Problem A bicycle rider is riding at a rate of 15 km/h for $\frac{4}{9}$ of his ride, 10 km/h for $\frac{2}{9}$ of his ride, and 8 km/h for the remainder of the ride. How much of his ride is he riding at a rate greater than 8 km/h?

SR-14 • Math Skill Handbook

Add and Subtract Fractions with Unlike Denominators To add or subtract fractions with unlike denominators, first find the least common denominator (LCD). This is the smallest number that is a common multiple of both denominators. Rename each fraction with the LCD, and then add or subtract. Find the simplest form if necessary.

Example 1

A chemist makes a paste that is $\frac{1}{2}$ table salt (NaCl), $\frac{1}{3}$ sugar ($C_6H_{12}O_6$), and the remainder is water (H_2O). How much of the paste is a solid?

Step 1 Find the LCD of the fractions.

$$\frac{1}{2} + \frac{1}{3} \text{ (LCD, 6)}$$

Step 2 Rename each numerator and each denominator with the LCD.

Step 3 Add the numerators.

$$\frac{3}{6} + \frac{2}{6} = \frac{(3+2)}{6} = \frac{5}{6}$$

$\frac{5}{6}$ of the paste is a solid.

Example 2

The average precipitation in Grand Junction, CO, is $\frac{7}{10}$ inch in November, and $\frac{3}{5}$ inch in December. What is the total average precipitation?

Step 1 Find the LCD of the fractions.

$$\frac{7}{10} + \frac{3}{5} \text{ (LCD, 10)}$$

Step 2 Rename each numerator and each denominator with the LCD.

Step 3 Add the numerators.

$$\frac{7}{10} + \frac{6}{10} = \frac{(7+6)}{10} = \frac{13}{10}$$

$\frac{13}{10}$ inches total precipitation, or $1\frac{3}{10}$ inches.

Practice Problem On an electric bill, about $\frac{1}{8}$ of the energy is from solar energy and about $\frac{1}{10}$ is from wind power. How much of the total bill is from solar energy and wind power combined?

Example 3

In your body, $\frac{7}{10}$ of your muscle contractions are involuntary (cardiac and smooth muscle tissue). Smooth muscle makes $\frac{3}{15}$ of your muscle contractions. How many of your muscle contractions are made by cardiac muscle?

Step 1 Find the LCD of the fractions.

$$\frac{7}{10} - \frac{3}{15} \text{ (LCD, 30)}$$

Step 2 Rename each numerator and each denominator with the LCD.

$$\frac{7 \times 3}{10 \times 3} = \frac{21}{30}$$

$$\frac{3 \times 2}{15 \times 2} = \frac{6}{30}$$

Step 3 Subtract the numerators.

$$\frac{21}{30} - \frac{6}{30} = \frac{(21-6)}{30} = \frac{15}{30}$$

Step 4 Find the GCF.

$$\frac{15}{30} \text{ (GCF, 15)}$$

$$\frac{1}{2}$$

$\frac{1}{2}$ of all muscle contractions are cardiac muscle.

Example 4

Tony wants to make cookies that call for $\frac{3}{4}$ of a cup of flour, but he only has $\frac{1}{3}$ of a cup. How much more flour does he need?

Step 1 Find the LCD of the fractions.

$$\frac{3}{4} - \frac{1}{3} \text{ (LCD, 12)}$$

Step 2 Rename each numerator and each denominator with the LCD.

$$\frac{3 \times 3}{4 \times 3} = \frac{9}{12}$$

$$\frac{1 \times 4}{3 \times 4} = \frac{4}{12}$$

Step 3 Subtract the numerators.

$$\frac{9}{12} - \frac{4}{12} = \frac{(9-4)}{12} = \frac{5}{12}$$

$\frac{5}{12}$ of a cup of flour

Practice Problem Using the information provided to you in Example 3 above, determine how many muscle contractions are voluntary (skeletal muscle).

Multiply Fractions To multiply with fractions, multiply the numerators and multiply the denominators. Find the simplest form if necessary.

> **Example**
>
> Multiply $\frac{3}{5}$ by $\frac{1}{3}$.
>
> **Step 1** Multiply the numerators and denominators.
>
> $$\frac{3}{5} \times \frac{1}{3} = \frac{(3 \times 1)}{(5 \times 3)} \; \frac{3}{15}$$
>
> **Step 2** Find the GCF.
>
> $\frac{3}{15}$ (GCF, 3)
>
> **Step 3** Divide the numerator and denominator by the GCF.
>
> $$\frac{3 \div 3}{15 \div 3} = \frac{1}{5}$$
>
> $\frac{3}{5}$ multiplied by $\frac{1}{3}$ is $\frac{1}{5}$.

Practice Problem Multiply $\frac{3}{14}$ by $\frac{5}{16}$.

Find a Reciprocal Two numbers whose product is 1 are called multiplicative inverses, or reciprocals.

> **Example**
>
> Find the reciprocal of $\frac{3}{8}$.
>
> **Step 1** Inverse the fraction by putting the denominator on top and the numerator on the bottom.
>
> $\frac{8}{3}$
>
> The reciprocal of $\frac{3}{8}$ is $\frac{8}{3}$.

Practice Problem Find the reciprocal of $\frac{4}{9}$.

Divide Fractions To divide one fraction by another fraction, multiply the dividend by the reciprocal of the divisor. Find the simplest form if necessary.

> **Example 1**
>
> Divide $\frac{1}{9}$ by $\frac{1}{3}$.
>
> **Step 1** Find the reciprocal of the divisor.
>
> The reciprocal of $\frac{1}{3}$ is $\frac{3}{1}$.
>
> **Step 2** Multiply the dividend by the reciprocal of the divisor.
>
> $$\frac{\frac{1}{9}}{\frac{1}{3}} = \frac{1}{9} \times \frac{3}{1} = \frac{(1 \times 3)}{(9 \times 1)} = \frac{3}{9}$$
>
> **Step 3** Find the GCF.
>
> $\frac{3}{9}$ (GCF, 3)
>
> **Step 4** Divide the numerator and denominator by the GCF.
>
> $$\frac{3 \div 3}{9 \div 3} = \frac{1}{3}$$
>
> $\frac{1}{9}$ divided by $\frac{1}{3}$ is $\frac{1}{3}$.

> **Example 2**
>
> Divide $\frac{3}{5}$ by $\frac{1}{4}$.
>
> **Step 1** Find the reciprocal of the divisor.
>
> The reciprocal of $\frac{1}{4}$ is $\frac{4}{1}$.
>
> **Step 2** Multiply the dividend by the reciprocal of the divisor.
>
> $$\frac{\frac{3}{5}}{\frac{1}{4}} = \frac{3}{5} \times \frac{4}{1} = \frac{(3 \times 4)}{(5 \times 1)} = \frac{12}{5}$$
>
> $\frac{3}{5}$ divided by $\frac{1}{4}$ is $\frac{12}{5}$ or $2\frac{2}{5}$.

Practice Problem Divide $\frac{3}{11}$ by $\frac{7}{10}$.

Use Ratios

When you compare two numbers by division, you are using a ratio. Ratios can be written 3 to 5, 3:5, or $\frac{3}{5}$. Ratios, like fractions, also can be written in simplest form.

Ratios can represent one type of probability, called odds. This is a ratio that compares the number of ways a certain outcome occurs to the number of possible outcomes. For example, if you flip a coin 100 times, what are the odds that it will come up heads? There are two possible outcomes, heads or tails, so the odds of coming up heads are 50:100. Another way to say this is that 50 out of 100 times the coin will come up heads. In its simplest form, the ratio is 1:2.

Example 1

A chemical solution contains 40 g of salt and 64 g of baking soda. What is the ratio of salt to baking soda as a fraction in simplest form?

Step 1 Write the ratio as a fraction.

$$\frac{salt}{baking\ soda} = \frac{40}{64}$$

Step 2 Express the fraction in simplest form. The GCF of 40 and 64 is 8.

$$\frac{40}{64} = \frac{40 \div 8}{64 \div 8} = \frac{5}{8}$$

The ratio of salt to baking soda in the sample is 5:8.

Example 2

Sean rolls a 6-sided die 6 times. What are the odds that the side with a 3 will show?

Step 1 Write the ratio as a fraction.

$$\frac{number\ of\ sides\ with\ a\ 3}{number\ of\ possible\ sides} = \frac{1}{6}$$

Step 2 Multiply by the number of attempts.

$$\frac{1}{6} \times 6\ attempts = \frac{6}{6}\ attempts = 1\ attempt$$

1 attempt out of 6 will show a 3.

Practice Problem Two metal rods measure 100 cm and 144 cm in length. What is the ratio of their lengths in simplest form?

Use Decimals

A fraction with a denominator that is a power of ten can be written as a decimal. For example, 0.27 means $\frac{27}{100}$. The decimal point separates the ones place from the tenths place.

Any fraction can be written as a decimal using division. For example, the fraction $\frac{5}{8}$ can be written as a decimal by dividing 5 by 8. Written as a decimal, it is 0.625.

Add or Subtract Decimals When adding and subtracting decimals, line up the decimal points before carrying out the operation.

Example 1

Find the sum of 47.68 and 7.80.

Step 1 Line up the decimal places when you write the numbers.

```
  47.68
+  7.80
```

Step 2 Add the decimals.

```
  1 1
  47.68
+  7.80
  55.48
```

The sum of 47.68 and 7.80 is 55.48.

Example 2

Find the difference of 42.17 and 15.85.

Step 1 Line up the decimal places when you write the number.

```
  42.17
- 15.85
```

Step 2 Subtract the decimals.

```
  3 11 1
  42.17
- 15.85
  26.32
```

The difference of 42.17 and 15.85 is 26.32.

Practice Problem Find the sum of 1.245 and 3.842.

Math Skill Handbook • **SR-17**

Multiply Decimals To multiply decimals, multiply the numbers like numbers without decimal points. Count the decimal places in each factor. The product will have the same number of decimal places as the sum of the decimal places in the factors.

Example

Multiply 2.4 by 5.9.

Step 1 Multiply the factors like two whole numbers.

$24 \times 59 = 1416$

Step 2 Find the sum of the number of decimal places in the factors. Each factor has one decimal place, for a sum of two decimal places.

Step 3 The product will have two decimal places.

14.16

The product of 2.4 and 5.9 is 14.16.

Practice Problem Multiply 4.6 by 2.2.

Divide Decimals When dividing decimals, change the divisor to a whole number. To do this, multiply both the divisor and the dividend by the same power of ten. Then place the decimal point in the quotient directly above the decimal point in the dividend. Then divide as you do with whole numbers.

Example

Divide 8.84 by 3.4.

Step 1 Multiply both factors by 10.

$3.4 \times 10 = 34$, $8.84 \times 10 = 88.4$

Step 2 Divide 88.4 by 34.

$$34 \overline{)88.4}$$
$$\,\,\underline{-68}$$
$$204$$
$$\underline{-204}$$
$$0$$

Quotient: 2.6

8.84 divided by 3.4 is 2.6.

Practice Problem Divide 75.6 by 3.6.

Use Proportions

An equation that shows that two ratios are equivalent is a proportion. The ratios $\frac{2}{4}$ and $\frac{5}{10}$ are equivalent, so they can be written as $\frac{2}{4} = \frac{5}{10}$. This equation is a proportion.

When two ratios form a proportion, the cross products are equal. To find the cross products in the proportion $\frac{2}{4} = \frac{5}{10}$, multiply the 2 and the 10, and the 4 and the 5. Therefore $2 \times 10 = 4 \times 5$, or $20 = 20$.

Because you know that both ratios are equal, you can use cross products to find a missing term in a proportion. This is known as solving the proportion.

Example

The heights of a tree and a pole are proportional to the lengths of their shadows. The tree casts a shadow of 24 m when a 6-m pole casts a shadow of 4 m. What is the height of the tree?

Step 1 Write a proportion.

$$\frac{\text{height of tree}}{\text{height of pole}} = \frac{\text{length of tree's shadow}}{\text{length of pole's shadow}}$$

Step 2 Substitute the known values into the proportion. Let h represent the unknown value, the height of the tree.

$$\frac{h}{6} \times \frac{24}{4}$$

Step 3 Find the cross products.

$h \times 4 = 6 \times 24$

Step 4 Simplify the equation.

$4h \times 144$

Step 5 Divide each side by 4.

$$\frac{4h}{4} \times \frac{144}{4}$$

$h = 36$

The height of the tree is 36 m.

Practice Problem The ratios of the weights of two objects on the Moon and on Earth are in proportion. A rock weighing 3 N on the Moon weighs 18 N on Earth. How much would a rock that weighs 5 N on the Moon weigh on Earth?

Use Percentages

The word *percent* means "out of one hundred." It is a ratio that compares a number to 100. Suppose you read that 77 percent of Earth's surface is covered by water. That is the same as reading that the fraction of Earth's surface covered by water is $\frac{77}{100}$. To express a fraction as a percent, first find the equivalent decimal for the fraction. Then, multiply the decimal by 100 and add the percent symbol.

Example 1

Express $\frac{13}{20}$ as a percent.

Step 1 Find the equivalent decimal for the fraction.

$$\begin{array}{r} 0.65 \\ 20\overline{)13.00} \\ \underline{12\ 0} \\ 1\ 00 \\ \underline{1\ 00} \\ 0 \end{array}$$

Step 2 Rewrite the fraction $\frac{13}{20}$ as 0.65.

Step 3 Multiply 0.65 by 100 and add the % symbol.

$$0.65 \times 100 = 65 = 65\%$$

So, $\frac{13}{20} = 65\%$.

This also can be solved as a proportion.

Example 2

Express $\frac{13}{20}$ as a percent.

Step 1 Write a proportion.

$$\frac{13}{20} = \frac{x}{100}$$

Step 2 Find the cross products.

$$1300 = 20x$$

Step 3 Divide each side by 20.

$$\frac{1300}{20} = \frac{20x}{20}$$
$$65\% = x$$

Practice Problem In one year, 73 of 365 days were rainy in one city. What percent of the days in that city were rainy?

Solve One-Step Equations

A statement that two expressions are equal is an equation. For example, $A = B$ is an equation that states that A is equal to B.

An equation is solved when a variable is replaced with a value that makes both sides of the equation equal. To make both sides equal the inverse operation is used. Addition and subtraction are inverses, and multiplication and division are inverses.

Example 1

Solve the equation $x - 10 = 35$.

Step 1 Find the solution by adding 10 to each side of the equation.

$$x - 10 = 35$$
$$x - 10 + 10 = 35 - 10$$
$$x = 45$$

Step 2 Check the solution.

$$x - 10 = 35$$
$$45 - 10 = 35$$
$$35 = 35$$

Both sides of the equation are equal, so $x = 45$.

Example 2

In the formula $a = bc$, find the value of c if $a = 20$ and $b = 2$.

Step 1 Rearrange the formula so the unknown value is by itself on one side of the equation by dividing both sides by b.

$$a = bc$$
$$\frac{a}{b} = \frac{bc}{b}$$
$$\frac{a}{b} = c$$

Step 2 Replace the variables a and b with the values that are given.

$$\frac{a}{b} = c$$
$$\frac{20}{2} = c$$
$$10 = c$$

Step 3 Check the solution.

$$a = bc$$
$$20 = 2 \times 10$$
$$20 = 20$$

Both sides of the equation are equal, so $c = 10$ is the solution when $a = 20$ and $b = 2$.

Practice Problem In the formula $h = gd$, find the value of d if $g = 12.3$ and $h = 17.4$.

Use Statistics

The branch of mathematics that deals with collecting, analyzing, and presenting data is statistics. In statistics, there are three common ways to summarize data with a single number—the mean, the median, and the mode.

The **mean** of a set of data is the arithmetic average. It is found by adding the numbers in the data set and dividing by the number of items in the set.

The **median** is the middle number in a set of data when the data are arranged in numerical order. If there were an even number of data points, the median would be the mean of the two middle numbers.

The **mode** of a set of data is the number or item that appears most often.

Another number that often is used to describe a set of data is the range. The **range** is the difference between the largest number and the smallest number in a set of data.

Example

The speeds (in m/s) for a race car during five different time trials are 39, 37, 44, 36, and 44.

To find the mean:

Step 1   Find the sum of the numbers.

$39 + 37 + 44 + 36 + 44 = 200$

Step 2   Divide the sum by the number of items, which is 5.

$200 \div 5 = 40$

The mean is 40 m/s.

To find the median:

Step 1   Arrange the measures from least to greatest.

36, 37, 39, 44, 44

Step 2   Determine the middle measure.

36, 37, <u>39</u>, 44, 44

The median is 39 m/s.

To find the mode:

Step 1   Group the numbers that are the same together.

44, 44, 36, 37, 39

Step 2   Determine the number that occurs most in the set.

<u>44, 44</u>, 36, 37, 39

The mode is 44 m/s.

To find the range:

Step 1   Arrange the measures from greatest to least.

44, 44, 39, 37, 36

Step 2   Determine the greatest and least measures in the set.

<u>44</u>, 44, 39, 37, <u>36</u>

Step 3   Find the difference between the greatest and least measures.

$44 - 36 = 8$

The range is 8 m/s.

Practice Problem Find the mean, median, mode, and range for the data set 8, 4, 12, 8, 11, 14, 16.

A **frequency table** shows how many times each piece of data occurs, usually in a survey. **Table 1** below shows the results of a student survey on favorite color.

Table 1 Student Color Choice		
Color	Tally	Frequency
red	IIII	4
blue	⋂⋂⋂⋂	5
black	II	2
green	III	3
purple	⋂⋂⋂⋂ II	7
yellow	⋂⋂⋂⋂ I	6

Based on the frequency table data, which color is the favorite?

SR-20 • Math Skill Handbook

Use Geometry

The branch of mathematics that deals with the measurement, properties, and relationships of points, lines, angles, surfaces, and solids is called geometry.

Perimeter The **perimeter** (P) is the distance around a geometric figure. To find the perimeter of a rectangle, add the length and width and multiply that sum by two, or $2(l + w)$. To find perimeters of irregular figures, add the length of the sides.

Example 1

Find the perimeter of a rectangle that is 3 m long and 5 m wide.

Step 1 You know that the perimeter is 2 times the sum of the width and length.

$P = 2(3\text{ m} + 5\text{ m})$

Step 2 Find the sum of the width and length.

$P = 2(8\text{ m})$

Step 3 Multiply by 2.

$P = 16\text{ m}$

The perimeter is 16 m.

Example 2

Find the perimeter of a shape with sides measuring 2 cm, 5 cm, 6 cm, 3 cm.

Step 1 You know that the perimeter is the sum of all the sides.

$P = 2 + 5 + 6 + 3$

Step 2 Find the sum of the sides.

$P = 2 + 5 + 6 + 3$

$P = 16$

The perimeter is 16 cm.

Practice Problem Find the perimeter of a rectangle with a length of 18 m and a width of 7 m.

Practice Problem Find the perimeter of a triangle measuring 1.6 cm by 2.4 cm by 2.4 cm.

Area of a Rectangle The **area** (A) is the number of square units needed to cover a surface. To find the area of a rectangle, multiply the length times the width, or $l \times w$. When finding area, the units also are multiplied. Area is given in square units.

Example

Find the area of a rectangle with a length of 1 cm and a width of 10 cm.

Step 1 You know that the area is the length multiplied by the width.

$A = (1\text{ cm} \times 10\text{ cm})$

Step 2 Multiply the length by the width. Also multiply the units.

$A = 10\text{ cm}^2$

The area is 10 cm².

Practice Problem Find the area of a square whose sides measure 4 m.

Area of a Triangle To find the area of a triangle, use the formula:

$A = \frac{1}{2}(\text{base} \times \text{height})$

The base of a triangle can be any of its sides. The height is the perpendicular distance from a base to the opposite endpoint, or vertex.

Example

Find the area of a triangle with a base of 18 m and a height of 7 m.

Step 1 You know that the area is $\frac{1}{2}$ the base times the height.

$A = \frac{1}{2}(18\text{ m} \times 7\text{ m})$

Step 2 Multiply $\frac{1}{2}$ by the product of 18×7. Multiply the units.

$A = \frac{1}{2}(126\text{ m}^2)$

$A = 63\text{ m}^2$

The area is 63 m².

Practice Problem Find the area of a triangle with a base of 27 cm and a height of 17 cm.

Circumference of a Circle The **diameter** (d) of a circle is the distance across the circle through its center, and the **radius** (r) is the distance from the center to any point on the circle. The radius is half of the diameter. The distance around the circle is called the **circumference** (C). The formula for finding the circumference is:

$$C = 2\pi r \text{ or } C = \pi d$$

The circumference divided by the diameter is always equal to 3.1415926... This nonterminating and nonrepeating number is represented by the Greek letter π (pi). An approximation often used for π is 3.14.

Example 1

Find the circumference of a circle with a radius of 3 m.

Step 1 You know the formula for the circumference is 2 times the radius times π.

$$C = 2\pi(3)$$

Step 2 Multiply 2 times the radius.

$$C = 6\pi$$

Step 3 Multiply by π.

$$C \approx 19 \text{ m}$$

The circumference is about 19 m.

Example 2

Find the circumference of a circle with a diameter of 24.0 cm.

Step 1 You know the formula for the circumference is the diameter times π.

$$C = \pi(24.0)$$

Step 2 Multiply the diameter by π.

$$C \approx 75.4 \text{ cm}$$

The circumference is about 75.4 cm.

Practice Problem Find the circumference of a circle with a radius of 19 cm.

Area of a Circle The formula for the area of a circle is: $A = \pi r^2$

Example 1

Find the area of a circle with a radius of 4.0 cm.

Step 1 $A = \pi(4.0)^2$

Step 2 Find the square of the radius.

$$A = 16\pi$$

Step 3 Multiply the square of the radius by π.

$$A \approx 50 \text{ cm}^2$$

The area of the circle is about 50 cm².

Example 2

Find the area of a circle with a radius of 225 m.

Step 1 $A = \pi(225)^2$

Step 2 Find the square of the radius.

$$A = 50625\pi$$

Step 3 Multiply the square of the radius by π.

$$A \approx 159043.1$$

The area of the circle is about 159043.1 m².

Example 3

Find the area of a circle whose diameter is 20.0 mm.

Step 1 Remember that the radius is half of the diameter.

$$A = \pi\left(\frac{20.0}{2}\right)^2$$

Step 2 Find the radius.

$$A = \pi(10.0)^2$$

Step 3 Find the square of the radius.

$$A = 100\pi$$

Step 4 Multiply the square of the radius by π.

$$A \approx 314 \text{ mm}^2$$

The area of the circle is about 314 mm².

Practice Problem Find the area of a circle with a radius of 16 m.

Volume The measure of space occupied by a solid is the **volume** (V). To find the volume of a rectangular solid multiply the length times width times height, or $V = l \times w \times h$. It is measured in cubic units, such as cubic centimeters (cm^3).

Example

Find the volume of a rectangular solid with a length of 2.0 m, a width of 4.0 m, and a height of 3.0 m.

Step 1 You know the formula for volume is the length times the width times the height.

$$V = 2.0 \text{ m} \times 4.0 \text{ m} \times 3.0 \text{ m}$$

Step 2 Multiply the length times the width times the height.

$$V = 24 \text{ m}^3$$

The volume is 24 m^3.

Practice Problem Find the volume of a rectangular solid that is 8 m long, 4 m wide, and 4 m high.

To find the volume of other solids, multiply the area of the base times the height.

Example 1

Find the volume of a solid that has a triangular base with a length of 8.0 m and a height of 7.0 m. The height of the entire solid is 15.0 m.

Step 1 You know that the base is a triangle, and the area of a triangle is $\frac{1}{2}$ the base times the height, and the volume is the area of the base times the height.

$$V = \left[\frac{1}{2}(b \times h)\right] \times 15$$

Step 2 Find the area of the base.

$$V = \left[\frac{1}{2}(8 \times 7)\right] \times 15$$

$$V = \left(\frac{1}{2} \times 56\right) \times 15$$

Step 3 Multiply the area of the base by the height of the solid.

$$V = 28 \times 15$$

$$V = 420 \text{ m}^3$$

The volume is 420 m^3.

Example 2

Find the volume of a cylinder that has a base with a radius of 12.0 cm, and a height of 21.0 cm.

Step 1 You know that the base is a circle, and the area of a circle is the square of the radius times π, and the volume is the area of the base times the height.

$$V = (\pi r^2) \times 21$$

$$V = (\pi 12^2) \times 21$$

Step 2 Find the area of the base.

$$V = 144\pi \times 21$$

$$V = 452 \times 21$$

Step 3 Multiply the area of the base by the height of the solid.

$$V \approx 9{,}500 \text{ cm}^3$$

The volume is about 9,500 cm^3.

Example 3

Find the volume of a cylinder that has a diameter of 15 mm and a height of 4.8 mm.

Step 1 You know that the base is a circle with an area equal to the square of the radius times π. The radius is one-half the diameter. The volume is the area of the base times the height.

$$V = (\pi r^2) \times 4.8$$

$$V = \left[\pi\left(\frac{1}{2} \times 15\right)^2\right] \times 4.8$$

$$V = (\pi 7.5^2) \times 4.8$$

Step 2 Find the area of the base.

$$V = 56.25\pi \times 4.8$$

$$V \approx 176.71 \times 4.8$$

Step 3 Multiply the area of the base by the height of the solid.

$$V \approx 848.2$$

The volume is about 848.2 mm^3.

Practice Problem Find the volume of a cylinder with a diameter of 7 cm in the base and a height of 16 cm.

Science Applications

Measure in SI

The metric system of measurement was developed in 1795. A modern form of the metric system, called the International System (SI), was adopted in 1960 and provides the standard measurements that all scientists around the world can understand.

The SI system is convenient because unit sizes vary by powers of 10. Prefixes are used to name units. Look at **Table 2** for some common SI prefixes and their meanings.

Table 2 Common SI Prefixes

Prefix	Symbol	Meaning	
kilo–	k	1,000	thousandth
hecto–	h	100	hundred
deka–	da	10	ten
deci–	d	0.1	tenth
centi–	c	0.01	hundreth
milli–	m	0.001	thousandth

Example

How many grams equal one kilogram?

Step 1 Find the prefix *kilo–* in **Table 2**.

Step 2 Using **Table 2**, determine the meaning of *kilo–*. According to the table, it means 1,000. When the prefix *kilo–* is added to a unit, it means that there are 1,000 of the units in a "kilounit."

Step 3 Apply the prefix to the units in the question. The units in the question are grams. There are 1,000 grams in a kilogram.

Practice Problem Is a milligram larger or smaller than a gram? How many of the smaller units equal one larger unit? What fraction of the larger unit does one smaller unit represent?

Dimensional Analysis

Convert SI Units In science, quantities such as length, mass, and time sometimes are measured using different units. A process called dimensional analysis can be used to change one unit of measure to another. This process involves multiplying your starting quantity and units by one or more conversion factors. A conversion factor is a ratio equal to one and can be made from any two equal quantities with different units. If 1,000 mL equal 1 L then two ratios can be made.

$$\frac{1{,}000 \text{ mL}}{1 \text{ L}} = \frac{1 \text{ L}}{1{,}000 \text{ mL}} = 1$$

One can convert between units in the SI system by using the equivalents in **Table 2** to make conversion factors.

Example

How many cm are in 4 m?

Step 1 Write conversion factors for the units given. From **Table 2**, you know that 100 cm = 1 m. The conversion factors are

$$\frac{100 \text{ cm}}{1 \text{ m}} \text{ and } \frac{1 \text{ m}}{100 \text{ cm}}$$

Step 2 Decide which conversion factor to use. Select the factor that has the units you are converting from (m) in the denominator and the units you are converting to (cm) in the numerator.

$$\frac{100 \text{ cm}}{1 \text{ m}}$$

Step 3 Multiply the starting quantity and units by the conversion factor. Cancel the starting units with the units in the denominator. There are 400 cm in 4 m.

$$4 \text{ m} = \frac{100 \text{ cm}}{1 \text{ m}} = 400 \text{ cm}$$

Practice Problem How many milligrams are in one kilogram? (Hint: You will need to use two conversion factors from **Table 2**.)

Table 3 Unit System Equivalents

Type of Measurement	Equivalent
Length	1 in = 2.54 cm 1 yd = 0.91 m 1 mi = 1.61 km
Mass and weight*	1 oz = 28.35 g 1 lb = 0.45 kg 1 ton (short) = 0.91 tonnes (metric tons) 1 lb = 4.45 N
Volume	1 in^3 = 16.39 cm^3 1 qt = 0.95 L 1 gal = 3.78 L
Area	1 in^2 = 6.45 cm^2 1 yd^2 = 0.83 m^2 1 mi^2 = 2.59 km^2 1 acre = 0.40 hectares
Temperature	°C = $\frac{(°F - 32)}{1.8}$ K = °C + 273

*Weight is measured in standard Earth gravity.

Convert Between Unit Systems Table 3 gives a list of equivalents that can be used to convert between English and SI units.

Example

If a meterstick has a length of 100 cm, how long is the meterstick in inches?

Step 1 Write the conversion factors for the units given. From **Table 3,** 1 in = 2.54 cm.

$$\frac{1 \text{ in}}{2.54 \text{ cm}} \text{ and } \frac{2.54 \text{ cm}}{1 \text{ in}}$$

Step 2 Determine which conversion factor to use. You are converting from cm to in. Use the conversion factor with cm on the bottom.

$$\frac{1 \text{ in}}{2.54 \text{ cm}}$$

Step 3 Multiply the starting quantity and units by the conversion factor. Cancel the starting units with the units in the denominator. Round your answer to the nearest tenth.

$$100 \text{ cm} \times \frac{1 \text{ in}}{2.54 \text{ cm}} = 39.37 \text{ in}$$

The meterstick is about 39.4 in long.

Practice Problem 1 A book has a mass of 5 lb. What is the mass of the book in kg?

Practice Problem 2 Use the equivalent for in and cm (1 in = 2.54 cm) to show how 1 in^3 ≈ 16.39 cm^3.

Precision and Significant Digits

When you make a measurement, the value you record depends on the precision of the measuring instrument. This precision is represented by the number of significant digits recorded in the measurement. When counting the number of significant digits, all digits are counted except zeros at the end of a number with no decimal point such as 2,050, and zeros at the beginning of a decimal such as 0.03020. When adding or subtracting numbers with different precision, round the answer to the smallest number of decimal places of any number in the sum or difference. When multiplying or dividing, the answer is rounded to the smallest number of significant digits of any number being multiplied or divided.

Example

The lengths 5.28 and 5.2 are measured in meters. Find the sum of these lengths and record your answer using the correct number of significant digits.

Step 1 Find the sum.

   5.28 m   2 digits after the decimal
 + 5.2 m   1 digit after the decimal
   10.48 m

Step 2 Round to one digit after the decimal because the least number of digits after the decimal of the numbers being added is 1.

The sum is 10.5 m.

Practice Problem 1 How many significant digits are in the measurement 7,071,301 m? How many significant digits are in the measurement 0.003010 g?

Practice Problem 2 Multiply 5.28 and 5.2 using the rule for multiplying and dividing. Record the answer using the correct number of significant digits.

Scientific Notation

Many times numbers used in science are very small or very large. Because these numbers are difficult to work with scientists use scientific notation. To write numbers in scientific notation, move the decimal point until only one non-zero digit remains on the left. Then count the number of places you moved the decimal point and use that number as a power of ten. For example, the average distance from the Sun to Mars is 227,800,000,000 m. In scientific notation, this distance is 2.278×10^{11} m. Because you moved the decimal point to the left, the number is a positive power of ten.

The mass of an electron is about 0.000 000 000 000 000 000 000 000 000 000 911 kg. Expressed in scientific notation, this mass is 9.11×10^{-31} kg. Because the decimal point was moved to the right, the number is a negative power of ten.

Example

Earth is 149,600,000 km from the Sun. Express this in scientific notation.

Step 1 Move the decimal point until one non-zero digit remains on the left.

   1.496 000 00

Step 2 Count the number of decimal places you have moved. In this case, eight.

Step 2 Show that number as a power of ten, 10^8.

Earth is 1.496×10^8 km from the Sun.

Practice Problem 1 How many significant digits are in 149,600,000 km? How many significant digits are in 1.496×10^8 km?

Practice Problem 2 Parts used in a high performance car must be measured to 7×10^{-6} m. Express this number as a decimal.

Practice Problem 3 A CD is spinning at 539 revolutions per minute. Express this number in scientific notation.

Make and Use Graphs

Data in tables can be displayed in a graph—a visual representation of data. Common graph types include line graphs, bar graphs, and circle graphs.

Line Graph A line graph shows a relationship between two variables that change continuously. The independent variable is changed and is plotted on the x-axis. The dependent variable is observed, and is plotted on the y-axis.

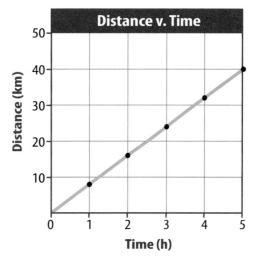

Figure 8 This line graph shows the relationship between distance and time during a bicycle ride.

Example

Draw a line graph of the data below from a cyclist in a long-distance race.

Table 4 Bicycle Race Data

Time (h)	Distance (km)
0	0
1	8
2	16
3	24
4	32
5	40

Step 1 Determine the x-axis and y-axis variables. Time varies independently of distance and is plotted on the x-axis. Distance is dependent on time and is plotted on the y-axis.

Step 2 Determine the scale of each axis. The x-axis data ranges from 0 to 5. The y-axis data ranges from 0 to 50.

Step 3 Using graph paper, draw and label the axes. Include units in the labels.

Step 4 Draw a point at the intersection of the time value on the x-axis and corresponding distance value on the y-axis. Connect the points and label the graph with a title, as shown in **Figure 8**.

Practice Problem A puppy's shoulder height is measured during the first year of her life. The following measurements were collected: (3 mo, 52 cm), (6 mo, 72 cm), (9 mo, 83 cm), (12 mo, 86 cm). Graph this data.

Find a Slope The slope of a straight line is the ratio of the vertical change, rise, to the horizontal change, run.

$$\text{Slope} = \frac{\text{vertical change (rise)}}{\text{horizontal change (run)}} = \frac{\text{change in } y}{\text{change in } x}$$

Example

Find the slope of the graph in **Figure 8**.

Step 1 You know that the slope is the change in y divided by the change in x.

$$\text{Slope} = \frac{\text{change in } y}{\text{change in } x}$$

Step 2 Determine the data points you will be using. For a straight line, choose the two sets of points that are the farthest apart.

$$\text{Slope} = \frac{(40 - 0) \text{ km}}{(5 - 0) \text{ h}}$$

Step 3 Find the change in y and x.

$$\text{Slope} = \frac{40 \text{ km}}{5 \text{ h}}$$

Step 4 Divide the change in y by the change in x.

$$\text{Slope} = \frac{8 \text{ km}}{\text{h}}$$

The slope of the graph is 8 km/h.

Math Skill Handbook • **SR-27**

Bar Graph To compare data that does not change continuously you might choose a bar graph. A bar graph uses bars to show the relationships between variables. The *x*-axis variable is divided into parts. The parts can be numbers such as years, or a category such as a type of animal. The *y*-axis is a number and increases continuously along the axis.

Example

A recycling center collects 4.0 kg of aluminum on Monday, 1.0 kg on Wednesday, and 2.0 kg on Friday. Create a bar graph of this data.

Step 1 Select the *x*-axis and *y*-axis variables. The measured numbers (the masses of aluminum) should be placed on the *y*-axis. The variable divided into parts (collection days) is placed on the *x*-axis.

Step 2 Create a graph grid like you would for a line graph. Include labels and units.

Step 3 For each measured number, draw a vertical bar above the *x*-axis value up to the *y*-axis value. For the first data point, draw a vertical bar above Monday up to 4.0 kg.

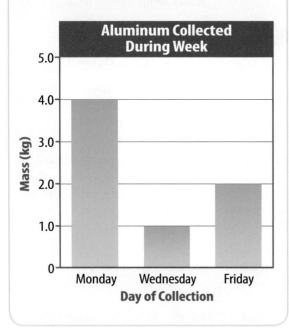

Practice Problem Draw a bar graph of the gases in air: 78% nitrogen, 21% oxygen, 1% other gases.

Circle Graph To display data as parts of a whole, you might use a circle graph. A circle graph is a circle divided into sections that represent the relative size of each piece of data. The entire circle represents 100%, half represents 50%, and so on.

Example

Air is made up of 78% nitrogen, 21% oxygen, and 1% other gases. Display the composition of air in a circle graph.

Step 1 Multiply each percent by 360° and divide by 100 to find the angle of each section in the circle.

$$78\% \times \frac{360°}{100} = 280.8°$$

$$21\% \times \frac{360°}{100} = 75.6°$$

$$1\% \times \frac{360°}{100} = 3.6°$$

Step 2 Use a compass to draw a circle and to mark the center of the circle. Draw a straight line from the center to the edge of the circle.

Step 3 Use a protractor and the angles you calculated to divide the circle into parts. Place the center of the protractor over the center of the circle and line the base of the protractor over the straight line.

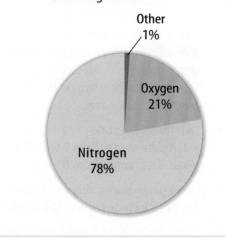

Practice Problem Draw a circle graph to represent the amount of aluminum collected during the week shown in the bar graph to the left.

FOLDABLES® Handbook

Student Study Guides & Instructions
By Dinah Zike

1. You will find suggestions for Study Guides, also known as Foldables or books, in each chapter lesson and as a final project. Look at the end of the chapter to determine the project format and glue the Foldables in place as you progress through the chapter lessons.

2. Creating the Foldables or books is simple and easy to do by using copy paper, art paper, and internet printouts. Photocopies of maps, diagrams, or your own illustrations may also be used for some of the Foldables. Notebook paper is the most common source of material for study guides and 83% of all Foldables are created from it. When folded to make books, notebook paper Foldables easily fit into 11" × 17" or 12" × 18" chapter projects with space left over. Foldables made using photocopy paper are slightly larger and they fit into Projects, but snugly. Use the least amount of glue, tape, and staples needed to assemble the Foldables.

3. Seven of the Foldables can be made using either small or large paper. When 11" × 17" or 12" × 18" paper is used, these become projects for housing smaller Foldables. Project format boxes are located within the instructions to remind you of this option.

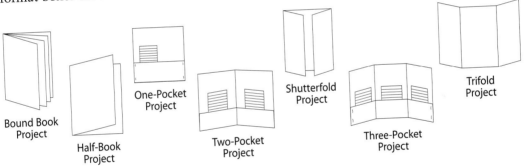

4. Use one-gallon self-locking plastic bags to store your projects. Place strips of two-inch clear tape along the left, long side of the bag and punch holes through the taped edge. Cut the bottom corners off the bag so it will not hold air. Store this Project Portfolio inside a three-hole binder. To store a large collection of project bags, use a giant laundry-soap box. Holes can be punched in some of the Foldable Projects so they can be stored in a three-hole binder without using a plastic bag. Punch holes in the pocket books before gluing or stapling the pocket.

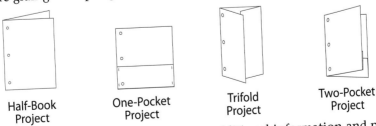

5. Maximize the use of the projects by collecting additional information and placing it on the back of the project and other unused spaces of the large Foldables.

Half-Book Foldable® By Dinah Zike

Step 1 Fold a sheet of notebook or copy paper in half.

Label the exterior tab and use the inside space to write information.

PROJECT FORMAT
Use 11" × 17" or 12" × 18" paper on the horizontal axis to make a large project book.

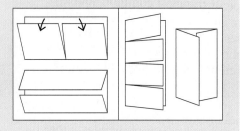

Variations
Paper can be folded horizontally, like a *hamburger* or vertically, like a *hot dog*.

C Half-books can be folded so that one side is ½ inch longer than the other side. A title or question can be written on the extended tab.

- -

Worksheet Foldable or Folded Book® By Dinah Zike

Step 1 Make a half-book (see above) using work sheets, internet print-outs, diagrams, or maps.

Step 2 Fold it in half again.

Variations

A This folded sheet as a small book with two pages can be used for comparing and contrasting, cause and effect, or other skills.

B When the sheet of paper is open, the four sections can be used separately or used collectively to show sequences or steps.

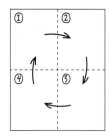

SR-30 • Foldables Handbook

Two-Tab and Concept-Map Foldable® By Dinah Zike

Step 1 Fold a sheet of notebook or copy paper in half vertically or horizontally.

Step 2 Fold it in half again, as shown.

Step 3 Unfold once and cut along the fold line or valley of the top flap to make two flaps.

Variations

A Concept maps can be made by leaving a ½ inch tab at the top when folding the paper in half. Use arrows and labels to relate topics to the primary concept.

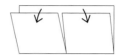

B Use two sheets of paper to make multiple page tab books. Glue or staple books together at the top fold.

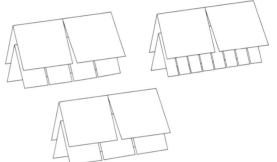

Three-Quarter Foldable® By Dinah Zike

Step 1 Make a two-tab book (see above) and cut the left tab off at the top of the fold line.

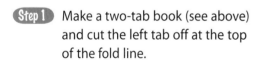

Variations

A Use this book to draw a diagram or a map on the exposed left tab. Write questions about the illustration on the top right tab and provide complete answers on the space under the tab.

B Compose a self-test using multiple choice answers for your questions. Include the correct answer with three wrong responses. The correct answers can be written on the back of the book or upside down on the bottom of the inside page.

Three-Tab Foldable® By Dinah Zike

Step 1 Fold a sheet of paper in half horizontally.

Step 2 Fold into thirds.

Step 3 Unfold and cut along the folds of the top flap to make three sections.

Variations

A Before cutting the three tabs draw a Venn diagram across the front of the book.

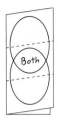

B Make a space to use for titles or concept maps by leaving a ½ inch tab at the top when folding the paper in half.

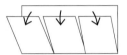

Four-Tab Foldable® By Dinah Zike

Step 1 Fold a sheet of paper in half horizontally.

Step 2 Fold in half and then fold each half as shown below.

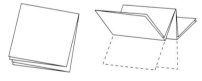

Step 3 Unfold and cut along the fold lines of the top flap to make four tabs.

Variations

A Make a space to use for titles or concept maps by leaving a ½ inch tab at the top when folding the paper in half.

B Use the book on the vertical axis, with or without an extended tab.

Folding Fifths for a Foldable® By Dinah Zike

Step 1 Fold a sheet of paper in half horizontally.

Step 2 Fold again so one-third of the paper is exposed and two-thirds are covered.

Step 3 Fold the two-thirds section in half.

Step 4 Fold the one-third section, a single thickness, backward to make a fold line.

Variations

A Unfold and cut along the fold lines to make five tabs.

B Make a five-tab book with a ½ inch tab at the top (see two-tab instructions).

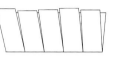

C Use 11" × 17" or 12" × 18" paper and fold into fifths for a five-column and/or row table or chart.

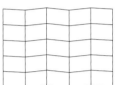

Folded Table or Chart, and Trifold Foldable® By Dinah Zike

Step 1 Fold a sheet of paper in the required number of vertical columns for the table or chart.

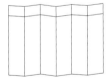

Step 2 Fold the horizontal rows needed for the table or chart.

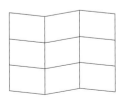

PROJECT FORMAT
Use 11" × 17" or 12" × 18" paper and fold it to make a large trifold project book or larger tables and charts.

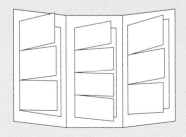

Variations

A Make a trifold by folding the paper into thirds vertically or horizontally.

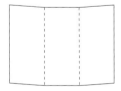

B Make a trifold book. Unfold it and draw a Venn diagram on the inside.

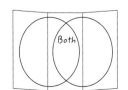

Foldables Handbook • SR-33

Two or Three-Pockets Foldable® By Dinah Zike

Step 1 Fold up the long side of a horizontal sheet of paper about 5 cm.

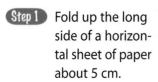

Step 2 Fold the paper in half.

Step 3 Open the paper and glue or staple the outer edges to make two compartments.

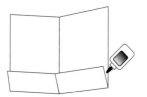

Variations

A Make a multi-page booklet by gluing several pocket books together.

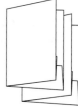

B Make a three-pocket book by using a trifold (see previous instructions).

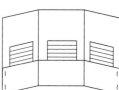

PROJECT FORMAT
Use 11" × 17" or 12" × 18" paper and fold it horizontally to make a large multi-pocket project.

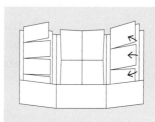

Matchbook Foldable® By Dinah Zike

Step 1 Fold a sheet of paper almost in half and make the back edge about 1–2 cm longer than the front edge.

Step 2 Find the midpoint of the shorter flap.

Step 3 Open the paper and cut the short side along the midpoint making two tabs.

Step 4 Close the book and fold the tab over the short side.

Variations

A Make a single-tab matchbook by skipping Steps 2 and 3.

B Make two smaller matchbooks by cutting the single-tab matchbook in half.

SR-34 • Foldables Handbook

Shutterfold Foldable® By Dinah Zike

Step 1 Begin as if you were folding a vertical sheet of paper in half, but instead of creasing the paper, pinch it to show the midpoint.

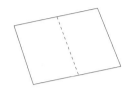

Step 2 Fold the top and bottom to the middle and crease the folds.

Variations

A Use the shutterfold on the horizontal axis.

B Create a center tab by leaving .5–2 cm between the flaps in Step 2.

PROJECT FORMAT
Use 11" × 17" or 12" × 18" paper and fold it to make a large shutterfold project.

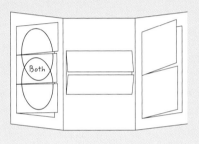

Four-Door Foldable® By Dinah Zike

Step 1 Make a shutterfold (see above).

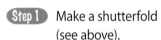

Step 2 Fold the sheet of paper in half.

Step 3 Open the last fold and cut along the inside fold lines to make four tabs.

Variations

A Use the four-door book on the opposite axis.

B Create a center tab by leaving .5–2 cm between the flaps in Step 1.

Bound Book Foldable® By Dinah Zike

Step 1 Fold three sheets of paper in half. Place the papers in a stack, leaving about .5 cm between each top fold. Mark all three sheets about 3 cm from the outer edges.

Step 2 Using two of the sheets, cut from the outer edges to the marked spots on each side. On the other sheet, cut between the marked spots.

Step 3 Take the two sheets from Step 1 and slide them through the cut in the third sheet to make a 12-page book.

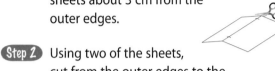

Step 4 Fold the bound pages in half to form a book.

Variation

A Use two sheets of paper to make an eight-page book, or increase the number of pages by using more than three sheets.

PROJECT FORMAT
Use two or more sheets of 11" × 17" or 12" × 18" paper and fold it to make a large bound book project.

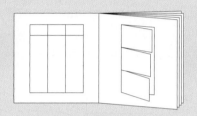

- -

Accordian Foldable® By Dinah Zike

Step 1 Fold the selected paper in half vertically, like a *hamburger*.

Step 2 Cut each sheet of folded paper in half along the fold lines.

Step 3 Fold each half-sheet almost in half, leaving a 2 cm tab at the top.

Step 4 Fold the top tab over the short side, then fold it in the opposite direction.

Variations

A Glue the straight edge of one paper inside the tab of another sheet. Leave a tab at the end of the book to add more pages.

B Tape the straight edge of one paper to the tab of another sheet, or just tape the straight edges of nonfolded paper end to end to make an accordian.

C Use whole sheets of paper to make a large accordian.

Layered Foldable® By Dinah Zike

Step 1 Stack two sheets of paper about 1–2 cm apart. Keep the right and left edges even.

Step 2 Fold up the bottom edges to form four tabs. Crease the fold to hold the tabs in place.

Step 3 Staple along the folded edge, or open and glue the papers together at the fold line.

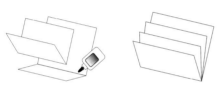

Variations

A Rotate the book so the fold is at the top or to the side.

B Extend the book by using more than two sheets of paper.

- -

Envelope Foldable® By Dinah Zike

Step 1 Fold a sheet of paper into a *taco*. Cut off the tab at the top.

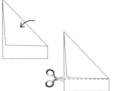

Step 2 Open the *taco* and fold it the opposite way making another *taco* and an X-fold pattern on the sheet of paper.

Step 3 Cut a map, illustration, or diagram to fit the inside of the envelope.

Step 4 Use the outside tabs for labels and inside tabs for writing information.

Variations

A Use 11" × 17" or 12" × 18" paper to make a large envelope.

B Cut off the points of the four tabs to make a window in the middle of the book.

Sentence Strip Foldable® By Dinah Zike

Step 1 Fold two sheets of paper in half vertically, like a *hamburger*.

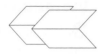

Step 2 Unfold and cut along fold lines making four half sheets.

Step 3 Fold each half sheet in half horizontally, like a *hot dog*.

Step 4 Stack folded horizontal sheets evenly and staple together on the left side.

Step 5 Open the top flap of the first sentence strip and make a cut about 2 cm from the stapled edge to the fold line. This forms a flap that can be raisied and lowered. Repeat this step for each sentence strip.

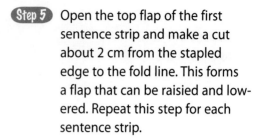

Variations

A Expand this book by using more than two sheets of paper.

B Use whole sheets of paper to make large books.

Pyramid Foldable® By Dinah Zike

Step 1 Fold a sheet of paper into a *taco*. Crease the fold line, but do not cut it off.

Step 2 Open the folded sheet and refold it like a *taco* in the opposite direction to create an X-fold pattern.

Step 3 Cut one fold line as shown, stopping at the center of the X-fold to make a flap.

Step 4 Outline the fold lines of the X-fold. Label the three front sections and use the inside spaces for notes. Use the tab for the title.

Step 5 Glue the tab into a project book or notebook. Use the space under the pyramid for other information.

Step 6 To display the pyramid, fold the flap under and secure with a paper clip, if needed.

Single-Pocket or One-Pocket Foldable® By Dinah Zike

Step 1 Using a large piece of paper on a vertical axis, fold the bottom edge of the paper upwards, about 5 cm.

Step 2 Glue or staple the outer edges to make a large pocket.

PROJECT FORMAT
Use 11" × 17" or 12" × 18" paper and fold it vertically or horizontally to make a large pocket project.

Variations

A Make the one-pocket project using the paper on the horizontal axis.

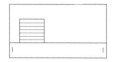

B To store materials securely inside, fold the top of the paper almost to the center, leaving about 2–4 cm between the paper edges. Slip the Foldables through the opening and under the top and bottom pockets.

Multi-Tab Foldable® By Dinah Zike

Step 1 Fold a sheet of notebook paper in half like a *hot dog*.

Step 2 Open the paper and on one side cut every third line. This makes ten tabs on wide ruled notebook paper and twelve tabs on college ruled.

Step 3 Label the tabs on the front side and use the inside space for definitions or other information.

Variation

A Make a tab for a title by folding the paper so the holes remain uncovered. This allows the notebook Foldable to be stored in a three-hole binder.

Reference Handbook

PERIODIC TABLE OF THE ELEMENTS

Element — Hydrogen
Atomic number — 1
Symbol — H
Atomic mass — 1.01
State of matter

- Gas
- Liquid
- Solid
- Synthetic

A column in the periodic table is called a **group**.

A row in the periodic table is called a **period**.

The number in parentheses is the mass number of the longest lived isotope for that element.

Period	1	2	3	4	5	6	7	8	9
1	Hydrogen 1 H 1.01								
2	Lithium 3 Li 6.94	Beryllium 4 Be 9.01							
3	Sodium 11 Na 22.99	Magnesium 12 Mg 24.31							
4	Potassium 19 K 39.10	Calcium 20 Ca 40.08	Scandium 21 Sc 44.96	Titanium 22 Ti 47.87	Vanadium 23 V 50.94	Chromium 24 Cr 52.00	Manganese 25 Mn 54.94	Iron 26 Fe 55.85	Cobalt 27 Co 58.93
5	Rubidium 37 Rb 85.47	Strontium 38 Sr 87.62	Yttrium 39 Y 88.91	Zirconium 40 Zr 91.22	Niobium 41 Nb 92.91	Molybdenum 42 Mo 95.96	Technetium 43 Tc (98)	Ruthenium 44 Ru 101.07	Rhodium 45 Rh 102.91
6	Cesium 55 Cs 132.91	Barium 56 Ba 137.33	Lanthanum 57 La 138.91	Hafnium 72 Hf 178.49	Tantalum 73 Ta 180.95	Tungsten 74 W 183.84	Rhenium 75 Re 186.21	Osmium 76 Os 190.23	Iridium 77 Ir 192.22
7	Francium 87 Fr (223)	Radium 88 Ra (226)	Actinium 89 Ac (227)	Rutherfordium 104 Rf (267)	Dubnium 105 Db (268)	Seaborgium 106 Sg (271)	Bohrium 107 Bh (272)	Hassium 108 Hs (270)	Meitnerium 109 Mt (276)

Lanthanide series

Cerium 58 Ce 140.12	Praseodymium 59 Pr 140.91	Neodymium 60 Nd 144.24	Promethium 61 Pm (145)	Samarium 62 Sm 150.36	Europium 63 Eu 151.96

Actinide series

Thorium 90 Th 232.04	Protactinium 91 Pa 231.04	Uranium 92 U 238.03	Neptunium 93 Np (237)	Plutonium 94 Pu (244)	Americium 95 Am (243)

					13	14	15	16	17	18
										Helium 2 He 4.00
					Boron 5 B 10.81	Carbon 6 C 12.01	Nitrogen 7 N 14.01	Oxygen 8 O 16.00	Fluorine 9 F 19.00	Neon 10 Ne 20.18
	10	11	12		Aluminum 13 Al 26.98	Silicon 14 Si 28.09	Phosphorus 15 P 30.97	Sulfur 16 S 32.07	Chlorine 17 Cl 35.45	Argon 18 Ar 39.95
	Nickel 28 Ni 58.69	Copper 29 Cu 63.55	Zinc 30 Zn 65.38	Gallium 31 Ga 69.72	Germanium 32 Ge 72.64	Arsenic 33 As 74.92	Selenium 34 Se 78.96	Bromine 35 Br 79.90	Krypton 36 Kr 83.80
	Palladium 46 Pd 106.42	Silver 47 Ag 107.87	Cadmium 48 Cd 112.41	Indium 49 In 114.82	Tin 50 Sn 118.71	Antimony 51 Sb 121.76	Tellurium 52 Te 127.60	Iodine 53 I 126.90	Xenon 54 Xe 131.29
	Platinum 78 Pt 195.08	Gold 79 Au 196.97	Mercury 80 Hg 200.59	Thallium 81 Tl 204.38	Lead 82 Pb 207.20	Bismuth 83 Bi 208.98	Polonium 84 Po (209)	Astatine 85 At (210)	Radon 86 Rn (222)
	Darmstadtium 110 Ds (281)	Roentgenium 111 Rg (280)	Copernicium 112 Cn (285)	Ununtrium * 113 Uut (284)	Ununquadium * 114 Uuq (289)	Ununpentium * 115 Uup (288)	Ununhexium * 116 Uuh (293)		Ununoctium * 118 Uuo (294)

* The names and symbols for elements 113-116 and 118 are temporary. Final names will be selected when the elements' discoveries are verified.

Gadolinium 64 Gd 157.25	Terbium 65 Tb 158.93	Dysprosium 66 Dy 162.50	Holmium 67 Ho 164.93	Erbium 68 Er 167.26	Thulium 69 Tm 168.93	Ytterbium 70 Yb 173.05	Lutetium 71 Lu 174.97
Curium 96 Cm (247)	Berkelium 97 Bk (247)	Californium 98 Cf (251)	Einsteinium 99 Es (252)	Fermium 100 Fm (257)	Mendelevium 101 Md (258)	Nobelium 102 No (259)	Lawrencium 103 Lr (262)

Metal
Metalloid
Nonmetal
Recently discovered

Glossary/Glosario

Multilingual eGlossary

A science multilingual glossary is available on the science Web site. The glossary includes the following languages.

Arabic	Hmong	Tagalog
Bengali	Korean	Urdu
Chinese	Portuguese	Vietnamese
English	Russian	
Haitian Creole	Spanish	

Cómo usar el glosario en español:
1. Busca el término en inglés que desees encontrar.
2. El término en español, junto con la definición, se encuentran en la columna de la derecha.

Pronunciation Key

Use the following key to help you sound out words in the glossary.

a	. . . .	back (BAK)	ew	. . . .	food (FEWD)
ay	. . . .	day (DAY)	yoo	. . . .	pure (PYOOR)
ah	. . . .	father (FAH thur)	yew	. . . .	few (FYEW)
ow	. . . .	flower (FLOW ur)	uh	. . . .	comma (CAH muh)
ar	. . . .	car (CAR)	u (+ con)	. . . .	rub (RUB)
e	. . . .	less (LES)	sh	. . . .	shelf (SHELF)
ee	. . . .	leaf (LEEF)	ch	. . . .	nature (NAY chur)
ih	. . . .	trip (TRIHP)	g	. . . .	gift (GIHFT)
i (i + com + e)	idea (i DEE uh)	j	. . . .	gem (JEM)
oh	. . . .	go (GOH)	ing	. . . .	sing (SING)
aw	. . . .	soft (SAWFT)	zh	. . . .	vision (VIH zhun)
or	. . . .	orbit (OR buht)	k	. . . .	cake (KAYK)
oy	. . . .	coin (COYN)	s	. . . .	seed, cent (SEED, SENT)
oo	. . . .	foot (FOOT)	z	. . . .	zone, raise (ZOHN, RAYZ)

English — A — Español

alkali (AL kuh li) metal/average atomic mass   **metal alcalino/masa atómica promedio**

alkali (AL kuh li) metal: an element in group 1 on the periodic table. (p. 357)

alkaline (AL kuh lun) earth metal: an element in group 2 on the periodic table. (p. 357)

atom: the smallest piece of an element that still represents that element. (p. 315)

atomic number: the number of protons in an atom of an element. (p. 327)

average atomic mass: the average mass of the element's isotopes, weighted according to the abundance of each isotope. (p. 329)

metal alcalino: elemento del grupo 1 de la tabla periódica. (pág. 357)

metal alcalinotérreo: elemento del grupo 2 de la tabla periódica. (pág. 357)

átomo: parte más pequeña de un elemento que mantiene la identidad de dicho elemento. (pág. 315)

número atómico: número de protones en el átomo de un elemento. (pág. 327)

masa atómica promedio: masa atómica promedio de los isótopos de un elemento, ponderado según la abundancia de cada isótopo. (pág. 329)

chemical bond/ionic bond   **enlace químico/enlace iónico**

C

chemical bond: a force that holds two or more atoms together. (p. 382)

chemical formula: a group of chemical symbols and numbers that represent the elements and the number of atoms of each element that make up a compound. (p. 394)

covalent bond: a chemical bond formed when two atoms share one or more pairs of valence electrons. (p. 391)

enlace químico: fuerza que mantiene unidos dos o más átomos. (pág. 382)

fórmula química: grupo de símbolos químicos y números que representan los elementos y el número de átomos de cada elemento que forman un compuesto. (pág. 394)

enlace covalente: enlace químico formado cuando dos átomos comparten uno o más pares de electrones de valencia. (pág. 391)

D

ductility (duk TIH luh tee): the ability to be pulled into thin wires. (p. 356)

ductilidad: capacidad para formar alambres delgados (pág. 356)

E

electron cloud: the region surrounding an atom's nucleus where one or more electrons are most likely to be found. (p. 322)

electron dot diagram: a model that represents valence electrons in an atom as dots around the element's chemical symbol. (p. 385)

electron: a negatively charged particle that occupies the space in an atom outside the nucleus. (p. 317)

nube de electrones: región que rodea el núcleo de un átomo en donde es más probable encontrar uno o más electrones. (pág. 322)

diagrama de puntos de Lewis: modelo que representa electrones de valencia en un átomo a manera de puntos alrededor del símbolo químico del elemento. (pág. 385)

electrón: partícula cargada negativamente que ocupa el espacio por fuera del núcleo de un átomo. (pág. 317)

G

group: a column on the periodic table. (p. 350)

grupo: columna en la tabla periódica. (pág. 350)

H

halogen (HA luh jun): an element in group 17 on the periodic table. (p. 365)

halógeno: elemento del grupo 17 de la tabla periódica. (pág. 365)

I

ion (I ahn): an atom that is no longer neutral because it has lost or gained valence electrons. (pp. 332, 398)

ionic bond: the attraction between positively and negatively charged ions in an ionic compound. (p. 400)

ión: átomo que no es neutro porque ha ganado o perdido electrones de valencia. (páges. 332, 398)

enlace iónico: atracción entre iones cargados positiva y negativamente en un compuesto iónico. (pág. 400)

isotopes/nucleus   **isótopos/núcleo**

isotopes: atoms of the same element that have different numbers of neutrons. (p. 328)

isótopos: átomos del mismo elemento que tienen diferente número de neutrones. (pág. 328)

luster: the way a mineral reflects or absorbs light at its surface. (p. 355)

brillo: forma en que un mineral refleja o absorbe la luz en su superficie. (pág. 355)

malleability (ma lee uh BIH luh tee): the ability of a substance to be hammered or rolled into sheets. (p. 356)

mass number: the sum of the number of protons and neutrons in an atom. (p. 328)

metal: an element that is generally shiny, is easily pulled into wires or hammered into thin sheets, and is a good conductor of electricity and thermal energy. (p. 355)

metallic bond: a bond formed when many metal atoms share their pooled valence electrons. (p. 401)

metalloid (MEH tul oyd): an element that has physical and chemical properties of both metals and nonmetals. (p. 367)

molecule (MAH lih kyewl): two or more atoms that are held together by covalent bonds and act as a unit. (p. 392)

maleabilidad: capacidad de una sustancia de martillarse o laminarse para formar hojas. (pág. 356)

número de masa: suma del número de protones y neutrones de un átomo. (pág. 328)

metal: elemento que generalmente es brillante, fácilmente puede estirarse para formar alambres o martillarse para formar hojas delgadas y es buen conductor de electricidad y energía térmica. (pág. 355)

enlace metálico: enlace formado cuando muchos átomos metálicos comparten su banco de electrones de valencia. (pág. 401)

metaloide: elemento que tiene las propiedades físicas y químicas de metales y no metales. (pág. 367)

molécula: dos o más átomos que están unidos mediante enlaces covalentes y actúan como una unidad. (pág. 392)

neutron: a neutral particle in the nucleus of an atom. (p. 321)

noble gas: an element in group 18 on the periodic table. (p. 366)

nonmetal: an element that has no metallic properties. (p. 363)

nuclear decay: a process that occurs when an unstable atomic nucleus changes into another more stable nucleus by emitting radiation. (p. 331)

nucleus: the region in the center of an atom where most of an atom's mass and positive charge are concentrated. (p. 320)

neutrón: partícula neutra en el núcleo de un átomo. (pág. 321)

gas noble: elemento del grupo 18 de la tabla periódica. (pág. 366)

no metal: elemento que tiene propiedades no metálicas. (pág. 363)

desintegración nuclear: proceso que ocurre cuando un núcleo atómico inestable cambia a otro núcleo atómico más estable mediante emisión de radiación. (pág. 331)

núcleo: región en el centro de un átomo donde se concentra la mayor cantidad de masa y las cargas positivas. (pág. 320)

period/valence electron   **periodo/electrón de valencia**

P

period: a row on the periodic table. (p. 350)

periodic table: a chart of the elements arranged into rows and columns according to their physical and chemical properties. (p. 345)

polar molecule: a molecule with a slight negative charge in one area and a slight positive charge in another area. (p. 393)

proton: positively charged particle in the nucleus of an atom. (p. 320)

periodo: hilera en la tabla periódica. (pág. 350)

tabla periódica: cuadro en que los elementos están organizados en hileras y columnas según sus propiedades físicas y químicas. (pág. 345)

molécula polar: molécula con carga ligeramente negativa en una parte y ligeramente positiva en otra. (pág. 393)

protón: partícula cargada positivamente en el núcleo de un átomo. (pág. 320)

R

radioactive: any element that spontaneously emits radiation. (p. 330)

radiactivo: cualquier elemento que emite radiación de manera espontánea. (pág. 330)

S

semiconductor: a substance that conducts electricity at high temperatures but not at low temperatures. (p. 367)

semiconductor: sustancia que conduce electricidad a altas temperaturas, pero no a bajas temperaturas. (pág. 367)

T

transition element: an element in groups 3–12 on the periodic table. (p. 358)

elemento de transición: elemento de los grupos 3–12 de la tabla periódica. (pág. 358)

V

valence electron: the outermost electron of an atom that participates in chemical bonding. (p. 384)

electrón de valencia: electrón más externo de un átomo que participa en el enlace químico. (pág. 384)

Index

Academic Vocabulary | *Italic numbers* = illustration/photo **Bold numbers** = vocabulary term
lab = indicates entry is used in a lab on this page

A

Academic Vocabulary, 330, 366, 402. *See also* **Vocabulary**
Actinide series, 359
Airship(s)
    green, 388
Alkali metal(s), *357,* **357**
Alkaline earth metal(s), **357**
Alpha decay, 331, *331*
Alpha particle(s)
    explanation of, 318, 331
    light created by, 319
    path of, *318,* 318–320, *319*
Argon
    atoms of, 385
    as noble gas, 366
Aristotle, 314
Atom(s)
    alpha particles and, 318, *318,* 319, *319*
    communicating your knowledge about, 334–335 *lab*
    early ideas about, 314, *314,* 317
    electrons in, 382–385, *383,* 385
    explanation of, *315,* **315,** 382
    mass number of, *328,* **328**
    neutral, 327, 332
    parts of, 326, *326*
    of polar and nonpolar molecules, *393,* 393
    size of, 315
    stable and unstable, 386, *386*
    technology to view, 315, *315*
    Thomson's model of, 317, *317,* 318
Atomic models
    Bohr's, 321, *321*
    continual change in, 324
    modern, 322, *322*
    path of, *318,* 319, *319*
    Rutherford's, 320, *320,* 321
    Thomson's, 317, *317,* 318
Atomic number
    of elements, 327
    explanation of, 347
Average atomic mass, *329,* **329**

B

Barium
    as alkaline earth metal, 357
Becquerel, Henri, 330, *330*
Beryllium
    as alkaline earth metal, 357
Beta decay, 331, *331*
Beta particles, 331
Big Idea, 310, 336, 342, 372, 378, 406
    Review, 339, 375, 409

Bohr, Neils, 321, *351*
Boiling point
    pattern in, 346
Bond(s). *See also* **Chemical bond(s)**
    explanation of, **390**
Boson(s), 324
Bromine
    properties of, 365, *365*

C

Cadmium
    on periodic table, 347, *347*
Calcium
    as alkaline earth metal, 357
Cancer
    treatments for, 324
Carbon
    in human body, 363, *363*
    isotopes of, 328, *328,* 329, *329*
    properties of, *364,* 365
Carbon dioxide
    chemical formula for, 394
Cathode ray tube(s), 316, *316*
Cathode ray(s)
    explanation of, 316, *316,* 317
    mass of, 317
Cesium
    as alkali metal, 357
Chadwick, James, 321
Chapter Review, 338–339, 374–375, 408–409
Chemical bond(s)
    covalent, *391,* 391–392, *392*
    explanation of, **382,** 383, 384, 386, **390**
Chemical formula(s), 394, **394**
Chemical reaction(s)
    limitations of, 330
Chlorine
    properties of, 365, *365*
Common Use. *See* **Science Use v. Common Use**
Compound(s)
    containing metals, 361
    covalent, 392–394, *393,* 394
    elements that make up, 390, 390 *lab*
    explanation of, **382**
    formation of, 394 *lab*
    halogens in, 365
    method to model, 396
Conduct, 402
Construct, 366
Copper
    on periodic table, 347, *347*
    uses for, 355
Covalent bond(s)
    double and triple, 392, *392*
    electron sharing and, 391

explanation of, **391,** *402*
Covalent compound(s)
    chemical compounds and, 394, *394*
    explanation of, 392
    molecule as unit of, 392
    nonpolar molecules and, 393, *393*
    polar molecules and, 393, *393*
Critical thinking, 352, 360, 369, 387, 395, 403, 409
Curie, Marie, 330, *330*
Curie, Pierre, 330

D

Dalton, John, 314, *314*
Democritus, 313, 314, *314*
Density
    explanation of, **356**
    of metals, 356
Ductility, 356

E

Electrode(s), 316, *316,* 317
Electron(s)
    bonding and, 383
    in Bohr's model, 321, *321*
    energy and, 383, *383*
    explanation of, **317,** 320
    mass of, 318
    in noble gas, 391
    noble gases and, 386, *386*
    number and arrangement of, 327, 332, 382, *383*
    properties of, 326, *326*
    shared, 391
    valence, 384, *384,* 386, 391, 392, *392,* 401
Electron cloud, 322, *322*
Electron dot diagram(s)
    explanation of, 385, *385,* 386
    function of, 396
Electron pooling, 401
Element(s). *See also specific elements*
    atomic number of, 327
    atoms of, 315
    explanation of, 327
    in periodic table, 345–350, *346, 347, 348–349, 350*
    synthetic, 351
    radioactive, 330, 331
    transition, *358,* 358–359
Energy
    electrons and, 383, *383*

F

Firework(s), 361
Fluorine
    properties of, 365, *365*

Fluorine ion, 332, *332*
Foldables, 317, 327, 337, 347, 356, 365, 369, 373, 383, 391, 398, 407
Francium
  as alkali metal, 357

G

Gamma decay, 331, *331*
Gamma ray(s), 331
Gas(es)
  noble, 366, 386, *386*
Germanium
  as semiconductor, 368
Gluons, 324
Gold
  on periodic table, 347, *347*
  properties of, 356
  uses for, 355, *356*
Gold foil experiment, 319, 320 *lab*
Green Science, 388
Group(s)
  in periodic table, **350,** 381, *382*
Gunpowder
  discovery of, 361

H

Halogen(s), 365, *365*
Helium
  atoms of, 385
  electron structure of, 386
  on periodic table, 349, *349*
Helium airship(s), 388
Higgs Boson, 324
Hindenburg, 388
Human beings
  elements in, 363, *363*
Hydrogen
  atoms of, 381, 385, 393
  in human body, 363, *363*
  properties of, 366
  in universe, 366
Hydrogen airship(s), 388
Hydrogen molecule(s), 393

I

Interpret Graphics, 352, 360, 369, 387, 395, 403
Iodine
  properties of, 365, *365*
Ion(s)
  determining charge of, 400
  explanation of, *332,* **332,** 398
  gaining valence electrons and, 399, *399*
  losing valence electrons and, 399, *399*
  in solution, 404–405 *lab*
Ionic bond(s), 400, *402*
Ionic compound(s)
  covalent compounds v., 401
  explanation of, 400
  ionic bonds in, 400
Isotope(s)
  explanation of, **328,** *328,* 329

  number of, 329, 329 *lab*
  radioactive, 331

K

Key Concepts, 344, 354, 362, 380, 389, 397
  Check, 344, 347, 350, 355, 356, 367, 383, 386, 390, 392, 401
  Summary, 372, 406
  Understand, 352, 360, 387, 395, 403, 408
Krypton
  as noble gas, 366

L

Lab, 334–335, 370–371, 404–405. *See also* **Launch Lab; MiniLab; Skill Practice**
Lanthanide series, 359
Launch Lab, 313, 326, 345, 355, 363, 381, 390, 398
Lepton(s), 324
Lesson Review, 323, 333, 352, 360, 369, 387, 395, 403
Lewis, Gilbert, 385
Lithium
  as alkali metal, 357, *357*
Luster, 355

M

Magnesium
  as alkaline earth metal, 357
Magnetic resonance imaging (MRI), 324, *324*
Malleability, 356, 359
Mass number, 328, 328
Math Skills, 328, 333, 339, 350, 352, 400, 403, 409
Matter
  early ideas about, *313,* 313–314
Melting point
  pattern in, 346
Mendeleev, Dimitri, 346, 347, 366
Mercury
  on periodic table, 347, *347*
Metal(s)
  alkali, 357, *357*
  alkaline earth, 357
  chemical properties of, 356
  explanation of, **355**
  patterns in properties of, 359, *359*
  in periodic table, 350, 355, 382
  physical properties of, 355–356, *356*
  as transition elements, 358, 358–359
  uses of, 355 *lab*
Metallic bond(s), 401, *402*
Metalloid(s)
  explanation of, **367**
  in periodic table, 350, 382
  properties and uses of, 368
  as semiconductors, 367
Mettner, Lise, *351*

Microscope(s)
  scanning tunneling, 315, *315*
MiniLab, 320, 329, 351, 359, 368, 386, 394, 401. *See also* **Lab**
Molecular model(s)
  function of, 394, *394*
Molecule(s)
  explanation of, **392**
  nonpolar, 393, *393*
  polar, 393, *393*
Moseley, Henry, 347

N

Negative ion(s), 332, *332*
Neon
  as noble gas, 366
  electron structure of, 386
Neutron(s)
  explanation of, **321,** 382, *382*
  properties of, 326, *326*
  quarks in, 322
Nitrogen
  compounds with, 365
  in human body, 363, *363*
Noble gas(es)
  electron arrangement in, 391
  explanation of, **366,** 386, *386*
Nonmetal(s)
  explanation of, **363**
  metals v., *364,* 364–365, *365*
  in periodic table, 350, 365, 382
  properties of, 363 *lab,* 364, *364*
Nonpolar molecule(s), 393, *393*
Nuclear decay, 331
Nucleus
  of atom, 382
  electron cloud around, 322, *322*
  explanation of, *320,* **320**

O

Oxygen
  atoms of, 391
  compounds with, 365
  in human body, 363, *363*

P

Period(s)
  in periodic table, **350,** 381, *382*
Periodic, 346
Periodic table. *See also specific elements*
  arrangement of, 353
  development of, *346,* 346–347, *347*
  element key in, 349, *349*
  explanation of, **345,** 346
  groups in, 350, *350,* 381, *382*
  illustration of, *348–349*
  modeling procedure used to develop, 370–371 *lab*
  organization of, 381 *lab,* 382
  periods in, 350, *350,* 381, *382*
  use by scientists of, 351

Phosphorus
compounds with, 365
in human body, 363, *363*
properties of, *364*
Photon(s), 324
Plutonium
use of, 359
Polar molecule(s), 393, *393*
Polonium, *330*
Positive ion(s), 332, *332*
Potassium
as alkali metal, 357, *357*
properties of, 359
Potassium nitrate, 361
Proton(s)
explanation of, **320,** 382, *382*
number of, 327, *327*, 328, 330
properties of, 326, *326*
quarks in, 322

Q

Quarks, 322

R

Radiation
explanation of, 330
types of, 331
Radiation therapy, 331
Radioactive isotope(s)
use of, 331
Radioactivity
discovery of, 330, *330*
Radium, *330*
as alkaline earth metal, 357
Radon
as noble gas, 366
Reading Check, 346, 351, 357, 359, 363, 365, 366, 368, 382, 385, 394, 398, 402
Review Vocabulary, 356, 382. *See also* **Vocabulary**
Rubidium
as alkali metal, 357
Rutherford, Ernest, 318–321

S

Saltpeter, 361
Sand
composition of, 313 *lab*

Scanning tunneling microscope (STM), 315, *315*
Science & Society, 324, 361
Science Methods, 335, 371, 405
Science Use v. Common Use, 348, 390. *See also* **Vocabulary**
Seaborg, Glen T., *351*
Selenium
properties of, 365
Semiconductor(s), 367
Silicon
explanation of, 367
uses of, 367, *367*, 368
Silver
on periodic table, 347, *347*
Skill Practice, 353, 396. *See also* **Lab**
Sodium
as alkali metal, 357, *357*
Sodium ion, 332, *332*
Spontaneous, 330
Standardized Test Practice, 340–341, 376–377, 410–411
Strontium
as alkaline earth metal, 357
Study Guide, 336–337, 372–373, 406–407
Subatomic particles, 324
Sugar
as covalent compound, 392
properties of, 393
Sulfur
in human body, 363, *363*
properties of, 365, 368

T

Thermal energy
conduction of, 359 *lab,* 368 *lab*
Thomson, J. J., 316–317
Transition element(s)
explanation of, **358**
lanthanide and actinide series, 359
properties of, 358
uses of, 358, *358*

U

Uranium, 330

V

Valence electron(s)
covalent bond and, 392, *392*
explanation of, **384,** *384*, 386, 401
metallic bonds and, 401
shared electrons as, 391
Visual Check, 349, 356, 364, 365, 384, 392
Vocabulary, 343, 344, 354, 362, 372, 379, 380, 389, 397, 406. *See also* **Academic Vocabulary; Review Vocabulary; Science Use v. Common Use; Word Origin**
Use, 352, 360, 369, 373, 387, 395, 403, 407

W

Water molecule
as polar, 393, *393*
What do you think?, 343, 352, 360, 369, 379, 387, 395, 403
Word Origin, 317, 328, 347, 356, 365, 367, 384, 393. *See also* **Vocabulary**
Writing In Science, 339, 375, 409

X

Xenon
as noble gas, 366

Z

Zinc
on periodic table, 347, *347*

Credits

Photo Credits

Cover Matt Meadows/Peter Arnold, Inc; ConnectED (t)Richard Hutchings, (c)Getty Images, (b)Jupiterimages/ThinkStock/Alamy; **ix** (b)Fancy Photography/Veer; **308** (t)Roger Ressmeyer/CORBIS, (c)Getty Images, (b)American Museum of Natural History; **310** (inset)Cern Photo/Frédéric Pitchal/Sygma/CORBIS; **310–311** (bkgd)James Brittain/Photolibrary; **312** (inset)Drs. Ali Yazdani & Daniel J. Hornbaker/Photo Researchers, Inc., (bkgd)Alessandro Della Bella/Keystone/CORBIS; **313** (t)Hutchings Photography/Digital Light Source, (l to r, t to b)Royalty-Free/CORBIS, (2)Creatas/PunchStock, (3)Royalty-Free/CORBIS, (4)DAJ/Getty Images; **314** (l)Scala/Art Resource, NY, (r)The Royal Institution, London, UK/The Bridgeman Art Library; **315** (t)Horizons Companies, (b)Drs. Ali Yazdani & Daniel J. Hornbaker/Photo Researchers, Inc.; **320** Hutchings Photography/Digital Light Source; **323** (t)Horizons Companies, (b)Drs. Ali Yazdani & Daniel J. Hornbaker/Photo Researchers, Inc.; **324** Royalty-Free/CORBIS; **325** (inset)Derrick Alderman/Alamy, (bkgd)Derrick Alderman/Alamy; **326** Hutchings Photography/Digital Light Source; **329** The McGraw-Hill Companies; **330** (t)SPL/Photo Researchers, Inc., (b)Time Life Pictures/Mansell/Time Life Pictures/Getty Images, (inset)figure 13 John Cancalosi/age fotostock; **334** (t to b)The McGraw-Hill Companies, (2)Hutchings Photography/Digital Light Source, (3)(4)The Mcgraw-Hill Companies; **335** Hutchings Photography/Digital Light Source; **339** James Brittain/Photolibrary; **342–343** Nick Caloyianis/National Geographics/Getty Images; **344** P.J. Stewart/Photo Researchers, Inc.; **345** Hutchings Photography/Digital Light Source; **347** (tl)DEA/A. Rizzi/De Agostini Picture Library/Getty Images, (tr)Astrid & Hanns-Frieder Michler/Photo Researchers, Inc., (cl)Visuals Unlimited/Ken Lucas/Getty Images, (cr)Richard Treptow/Photo Researchers, Inc., (bl)CORBIS, (br)ImageState/Alamy; **350** (l)David J. Green/Alamy, (c)Wildlife/Peter Arnold Inc., (r)Mark Schneider/Visuals Unlimited/Getty Images; **351** (l)LBNL/Photo Researchers, Inc., (c)Boyer/Roger Viollet/Getty Images, (r)Ullstein Bild/Peter Arnold, Inc.; **353** Hutchings Photography/Digital Light Source; **354** Paul Katz/photolibrary.com; **355** Hutchings Photography/Digital Light Source; **356** (tl)The McGraw-Hill Companies, (tc)Paul Katz/Getty Images, (tr)Egyptian National Museum, Cairo, Egypt, Photo © Boltin Picture Library/The Bridgeman Art Library International; **356** (bl)NASA, (bc)Hutchings Photography/Digital Light Source, (br)Charles Stirling/Alamy; **357** (l)The McGraw-Hill Companies, Inc./Stephen Frisch, photographer, (c)sciencephotos/Alamy, (r)Martyn Chillmaid/Oxford Scientific (OSF)/photolibrary.com; **358** (l)Royalty-Free/CORBIS, (cl)Dr. Parvinder Sethi, (cr)Joel Arem/Photo Researchers, Inc., (r)Ingram Publishing/SuperStock; **359** Hutchings Photography/Digital Light Source; **360** (t)Egyptian National Museum, Cairo, Egypt, Photo © Boltin Picture Library/The Bridgeman Art Library International, (c)The McGraw-Hill Companies, Inc./Stephen Frisch, photographer, (b)Paul Katz/Getty Images; **361** Jeff Hunter/Getty Images; **362** E.O. lawrence Berkely National Laboratory, University of California, U.S. Department of Energy; **363** Hutchings Photography/Digital Light Source; **364** (tl)Ted Foxx/Alamy (tr)Richard Treptow/Photo Researchers, Inc., (c)Hutchings Photography/Digital Light Source, (bl)Photodisc/Getty Images, (br)Charles D. Winters/Photo Researchers, Inc.; **365** sciencephotos/Alamy; **366** NASA-JPL; **367** (l)Ingemar Aourell/Getty Images, (cl)Don Farrall/Getty Images, (cr)Gabe Palmer/Alamy, (r)Henrik Sorensen/Getty Images; **368** (t)PhotoLink/Getty Images, (b)Hutchings Photography/Digital Light Source; **369** (t)Richard Treptow/Photo Researchers, Inc., (c)sciencephotos/Alamy, (b)PhotoLink/Getty Images; **370 371** (b)Hutchings Photography/Digital Light Source; **372** (t)David J. Green/Alamy, (b)Mark Schneider/Visuals Unlimited/Getty Images; **375** Nick Caloyianis/National Geographics/Getty Images; **378–379** altrendo images/Getty Images; **379** Hutchings Photography/Digital Light Source; **380** Douglas Fisher/Alamy; **381 382 386** Hutchings Photography/Digital Light Source; **388** (t)Popperfoto/Getty Images, (c)Underwood & Underwood/CORBIS, (bl)John Meyer, (br)Ilene MacDonald/Alamy; **389** Gazimal/Getty Images; **390 394** Hutchings Photography/Digital Light Source; **396** (l)Macmillan/McGraw-Hill, (r)Hutchings Photography/Digital Light Source; **397** Brent Winebrenner/Photolibrary.com; **398** Hutchings Photography/Digital Light Source; **402** (t)Photodisc/Getty Images, (c)C Squared Studios/Getty Images, (b)Jennifer Martine/Jupiter Images; **404** (tr)(cr) (bl)Macmillan/McGraw-Hill, (br)Hutchings Photography/Digital Light Source, (t to b)(2)Hutchings Photography/Digital Light Source, (3)Macmillan/McGraw-Hill, (4) Hutchings Photography/Digital Light Source, (5)(6)Macmillan/McGraw-Hill, (7)(8)Richard Hutchings (see Digital Light Source), (9)Macmillan/McGraw-Hill; **405** Hutchings Photography/Digital Light Source; **409** altrendo images/Getty Images; **SR-0–SR-1** (bkgd)Gallo Images—Neil Overy/Getty Images; **SR-2** Hutchings Photography/Digital Light Source; **SR-6** Michell D. Bridwell/PhotoEdit; **SR-7** (t)The McGraw-Hill Companies, (b)Dominic Oldershaw; **SR-8** StudiOhio; **SR-9** Timothy Fuller; **SR-10** Aaron Haupt; **SR-12** KS Studios; **SR-13 SR-47** Matt Meadows; **SR-48** (c)NIBSC/Photo Researchers, Inc., (r)Science VU/Drs. D.T. John & T.B. Cole/Visuals Unlimited, Inc, Stephen Durr; **SR-49** (t)Mark Steinmetz, (r)Andrew Syred/Science Photo Library/Photo Researchers, (br)Rich Brommer; **SR-50** (l)Lynn Keddie/Photolibrary, (tr)G.R. Roberts, David Fleetham/Visuals Unlimited/Getty Images; **SR-51** Gallo Images/CORBIS.